Jerome Van Crowninshield Smith

**The Ways of Women in their Physical, Moral and Intellectual**

**Relations**

Jerome Van Crowninshield Smith

**The Ways of Women in their Physical, Moral and Intellectual Relations**

ISBN/EAN: 9783337365769

Printed in Europe, USA, Canada, Australia, Japan

Cover: Foto ©berggeist007 / pixelio.de

More available books at **www.hansebooks.com**

# THE

# WAYS OF WOMEN

IN THEIR

## PHYSICAL, MORAL

AND

## INTELLECTUAL RELATIONS.

BY A MEDICAL MAN.

———

NEW YORK:

JOHN P. JEWETT & CO., PUBLISHERS,

No. 5 DEY STREET.

1873.

LANGE, LITTLE & HILLMAN,
PRINTERS, ELECTROTYPERS AND STEREOTYPERS,
108 TO 114 WOOSTER STREET, N. Y.

# PREFACE.

A LEADING object of this volume is to explain, in a familiar manner, how women may improve their condition by conforming to the laws of health.

Next, to point out the way by which, in this active age of Christian civilization, they may be qualified for sustaining themselves honorably and successfully in various new relations to society.

# CONTENTS.

# CHAPTER IX.

# CHAPTER X.

# CHAPTER XI.

# CHAPTER XII.

# CHAPTER XIII.

# CHAPTER XIV.

8 CONTENTS.

# The Ways of Women.

## CHAPTER I.

Their Influence, Magnetism, Inborn Intuitions, and Power in every Age and Country.

SINCE the creation of Eve, women have been objects of peculiar interest wherever seen. They are conscious of possessing a controlling influence over men, whatever their social position, and they wield it according to circumstances. They assume a general attitude of defence, as though recognizing the fact of being physically weak, while exercising a mysterious strength which no man has the energy to resist. Whatever her condition, from a pampered lady of the court to a menial servant of the kitchen, every woman demonstrates in her intercourse with the world the truth of the foregoing proposition. Her attractions or exhibitions of contempt are acts of volition. Both may be exerted either for good or for evil, according to her own individual determination.

There are peculiar inborn properties of the sex which education modifies but cannot extinguish. Beauty, elegance of form, and grace of manners are powerful auxiliary forces when exercised for the accomplishment of ambitious designs. There is neither spirit nor persistency enough in the whole range of masculine humanity, with but a few rare exceptions, to withstand the artillery of a magnificent woman's charms, when sent forth in all their potency with a view to conquest.

Kings, princes, statesmen, theologians, and those of grave and solemn deportment, are alike impressible when subjected to those mysterious influences which are the glory and the shame of womanhood. Science sheds no light on this subject, since it has not yet been explained how female organization is endowed with such superior force.

In the functions of organs essential to nutrition, and in the form and offices of the apparatus of the special senses, there is no apparent difference, and yet men and women differ in their natures. Neither one is a perfect being. They are complete halves. The two constitute one perfect whole.

## The Frame of Woman.

There are about two hundred and forty-eight bones in a human skeleton. More are often found, but fewer than two hundred and thirty-nine could not be dispensed with, and the individual not be noticeable as organically defective.

When extra bones appear, they are usually under the balls of the great toes. From their resemblance to sesamum seeds, they are called sesamoids.

The production of those split-pea shaped bones may happen at any period of life about the articulations of the thumbs, fingers, or toes, to meet certain contingencies to which they may have been exposed. Their development under flexor tendons are purely a mechanical principle, to carry the cord farther from the joint to increase its power. In some cases the introduction of those extra bones is a temporary provision, and they are absorbed and taken away when no longer of service.

The knee-pans are of the same character, being movable

fulcrums, rising and falling in the flexion or extension of the limb. By placing the palm of the hand over the knee-cap while bending the leg, the sliding motion of the patella, up and down, illustrates its office in the economy of that joint.

When extra burdens are imposed for a succession of weeks or months, requiring a firmer foothold in order to carry the weight steadily, the cordage of the feet will increase, both in volume and tone, to meet the emergency. Thus, a hodcarrier, climbing ladders, will not only have enlarged feet, but sesamoid bones make their appearance at points where the tendons have the greatest amount of strain upon them, about the under side of the toes.

## Laws to Meet Cases.

Nature exercises a discretionary oversight, as it were, for the comfort of the individual as well as for the immediate protection of a most exposed part, by introducing temporary assistance, and removing it when no longer necessary.

Small ossific deposits sometimes appear about the finger joints, for the same beneficial purpose. Should they become inconveniently large, when the cause is removed which quickened them into existence, ordinarily they begin to diminish in size, unless the individual is at an advanced age, when vitality loses much of its former force.

## Equal Number of Bones in both Sexes.

There are exactly as many bones in the female as in the male skeleton, but they are smaller and more delicate in texture, with slighter depressions and less prominent eminences upon them. A female skull is smaller, thinner, and bears upon its general exterior, peculiarities indicative of mental qualities, if there is any reliance upon phrenology, not to be neglected or overlooked in studying osteological architecture.

One of the most obvious differences is a gentle arching of the female head from the forehead upward to the vertex, which is always more elevated than in male skulls. Scarcely one flat head, in that region, can be found in a thousand. On the contrary, the number of upwardly arched heads is small, in comparison, among men. They are more commonly quite flat, or slightly raised between the sinciput and occiput.

This characteristic difference is considered, by experts in sentimental craniology, as proof positive, that women always have more elevated moral sentiment, and are actually better than men, because they possess a more favorable organization. Nothing is more familiar than bones, and therefore, little or no thought is bestowed upon them. But, when carefully examined, they are rich in lessons of instruction. They are levers for the attachment of muscles or movers, by the contraction or relaxation of which motions are effected.

Every animal which is capable of making a motion possesses muscles. Most of them have skeletons clothed with flesh, and that is an aggregation of muscles. In the simpler forms of aquatic life, as in lobsters, crabs, etc., etc., the skeleton is on the outside. While it gives attachment to muscles, it also is a coat of mail, a house or a fortress in which they dwell, securely defended from the assaults of enemies.

### FORMATION OF BONES.

At birth we have no perfect bones, with the exception of the auditory, but they soon begin to harden as the infant is furnished with food. Then ossification commences,—a very gradual process, not fairly completed till about the twentieth year.

The formation, therefore, of two hundred and forty-eight hard bones of different shapes, densities, and positions, out of food taken into the stomach, is a marvel. But that is not the whole of the wonder. When fashioned and apparently finished, then they are taken to pieces, particle by particle, and carried out of the body, a new particle invariably being inserted when an old one is removed.

There is no cessation of this vital process; it is perpetually going on from the hour of birth to the expiration of the last breath. It is not unlike building a brick edifice. When completed, were the masons to commence forthwith to remove a brick in the wall, and, at the same instant, introduce a new one in its place, and never relax in that repetition of exchanging new for old ones, till the structure was destroyed, it would represent the process always going on in a living being.

Our very bones are many times renewed, therefore, in the course of a medium lifetime, although their composition is a compound of phosphate of magnesia, phosphate and carbonate of lime, manganese, iron, silex, etc., in definite proportions, which no chemist could more accurately weigh in his scales.

## MALE AND FEMALE SKELETON.

Although constructed of exactly the same materials, in the same elementary proportions, having the same general forms, there is a difference in the skeletons which the anatomist detects very readily.

When suspended side by side, a characteristic difference becomes apparent. The pelvis is broader and deeper in the female, which throws the hips further apart, giving to that central pivot of the frame a feature which artists, particularly sculptors, are careful to note, as it cants the knees so nearly together they almost touch. In the male subject the thigh bones are nearly parallel.

Again, the necks of the femoral bones are longer in the female, throwing the shafts further from the sockets in which they are articulated. A vertical line drawn perpendicularly to the space between the knees, from the chin, gives the most satisfactory demonstration of this very curious arrangement.

The distance between the articulating heads of the thigh bones is so plainly recognized, as to enable those with a very limited acquaintance with osteology, to determine, with considerable accuracy, to which sex a skeleton belonged.

This circumstance may be of considerable importance in conducting judicial inquiries. Public anxiety is sometimes painfully excited when human remains are found in obscure places, that lead to the suspicion of a concealed crime. If a man had mysteriously disappeared, and the discovered bones belonged to a female, it would be important in settling a mooted question.

# CHAPTER II.

## GENERALIZATIONS.

References in Construction to Specific Purposes—Rudimentary Organs—
Constant Evidences of Design—Organic Life and Multiform Objects of
Interest in the Investigation of Laws Regulating Existence.

FROM the beginning of woman's existence, a reference is discoverable in her mind and body, in regard to the exact position she was predestined to occupy. As already expressed, her bones, not in their composition, but in some of their directions rather than in their forms, indicate a reason for deviations from lines given to those of the male. They must have had the same condition in the first created woman, otherwise the architecture manifested in the pelvic construction would have been an imperfection. Eve would have left no posterity on the earth, had the carpentry of that region been different from what it now is in her feminine descendants.

Small philosophers have dared to suggest that Adam was, in his own person, both male and female. Rudimentary paps of men, monkies, dogs, swine, and many other quadrupeds, are cited as testifying to the truth of their theory, that they were originally hermaphrodites—being in their present state, substantially, degenerated females. In the first chapter of Genesis, they find a declaration respecting the first man, which strengthens their convictions.

Woman, then, in the peculiarities of her bones, presents evidences of a design which could not have been so without a designer. She did not fashion herself; and, therefore, in the commencement of our inquiries, are irrefragable proofs of a Supreme Intelligence in every step of progress in these investigations.

The sexes have been distinct from the beginning.

It would be inappropriate to discuss the mental attributes of women in the commencement of these deliberations, or to institute comparisons by weighing her brain in patent balances, measuring the length of her muscles, counting the hairs on her head, or drawing parallels between her attainments and those of giants in art, literature, and science. It is necessary to keep within prescribed boundaries in order to gain accurate knowledge of her ways, by studying carefully what is already known, to find out what may be unknown, that would enhance her claims for better treatment and justice at the hands of those who are her natural protectors and associates.

By contraction, muscles bend the arm, raise a shoulder, grit the teeth, or carry a spoon to the mouth. There are no surprises excited by motions so common. Women walk, run, eat, drink, sleep, and recruit their exhausted vitality as men do. By analagous mechanism they perform on musical instruments; think, speak, sing, and express their sensations. Therefore, their brain is the same in form and texture, but smaller, and hence it has been hastily concluded they are unequal to enterprises in which men excel. Only those quite incompetent to comprehend the mission of women, or appreciate her many claims to distinction, arrive at that conclusion.

### What they Have Done.

By hereditary right women have ascended thrones. History narrates thrilling military successes of women. In strategy, they excel when they choose to exercise their ingenuity.

They have risen to an enviable distinction without wealth,

by the practice of several arts, and also by varied intellectual attainments. Aside from the immense aid of personal charms, which a few of the many make stepping-stones to eminent positions, their bravery, heroism, and indomitable perseverance have always been themes for admiration, which poets and historians seize upon with avidity for illustrating their capacity and their eminent success in all ages.

They struggle mightily and die valiantly in defence of their honor. Guns, swords, batteries, armies, and ships-of-war are set in motion by men for the subjugation of an enemy. Women bring conquerors to their feet with the magic of their eyes.

### PHILOSOPHICAL REFLECTIONS.

Although osteology has been referred to, a study of the bones of all kinds of animals, it would be a profitable study in female schools and seminaries. There is nothing improper, revolting, frightful, or disgusting in the pursuit. No better opportunity ever presents for impressing upon the plastic minds of youth, properly presented, an overwhelming argument against infidelity, than a plain demonstration of the skill and superhuman contrivances exhibited in the adjustment of the bones of a bird, a carnivorous beast, or, better still, in the construction of a human skull.

Children must see things to understand them. The eye takes in a group at once, and the impressions made by tangible illustrations of the resources of the Divine Originator, in the examination of such mechanism, cannot be easily forgotten. To see the tubular bridge spanning the Straits of Menai, the traveller has ever after a vivid recollection of its appearance and utility, which he could not have by simply reading about it. Anatomical researches fail to show any very striking differences in the general construction of men and women. Bones, muscles, nerves, blood vessels, most of the glands and viscera, are precisely alike in shape and function.

It is not enough to state explicitly, that all the internal apparatus of organic life so much resemble each other when detached from the cavities in which they were lodged the most experienced student of a dissecting-room could not decide which were taken from a male, or which from a female. Products of secreting glands, as the salivary in the mouth; the lachrymal in the orbits; wax in the external ear, etc., are precisely the same in composition. In short, whatever is necessary for sustaining life in the one is equally so in the other, and accomplished precisely in the same manner.

These generalizations are neither new nor equally interesting to all; nevertheless, they are curious facts, and not unworthy the thoughtful consideration of those who confess their belief in the existence of a Being who alone could have originated these complicated mechanisms, and established laws which secure for them, as they do for planets in their orbits, perfect harmony in their movements.

### Nothing by Chance.

Two deviations in the bones of the female have been special points of interest, not on account of their texture or relations, but because they indicate, unmistakably, an office which the same bones in a man were not to have.

The collar-bones, or clavicles, are invariably longer in women than in men. Whether she is short or tall, those bones always maintain the observable proportionable length to the rest of the skeleton; otherwise, there are no peculiarities. Attachments of ligaments, muscles, the course of vessels over or under them, are no way different in the sexes. While women inhabit the earth, those collar-bones will have the same relative length, whether mothers nurse their babes as they should or not.

By their extra elongation, their shoulder-blades are forced farther back towards the spine, thus making a broader flooring, or space, for the lodgment of the breasts in front. This is a reason why women cannot exercise their arms gracefully in throwing a ball. Rarely, indeed, can they hit a mark in that exercise, even with hours of practice. Their awkwardness in that respect is proverbial; not, however, from any neglect in the education of the muscles of the arm, but from a congenital conformation, are they less expert than men in throwing. The difficulty lies in the arrangement of the ends of the muscles, further removed from the shoulder-joint, by reason of longer clavicles.

## VOCAL BOX.

That protuberance of the upper part of the throat, vulgarly called Adam's Apple, from a tradition that the forbidden fruit stuck there—described in books under the name of larynx, or vocal box—is a genuine musical instrument. Within it there are vocal chords, which vibrate as the current of air passes their thin edges. The sound thus produced is voice, afterwards modulated, and by systematic practice forms a language.

In men that box, at puberty, becomes enlarged and partially ossifies. At that period of development the boy's voice is irregular—a *vox rauca*—a sign that he is passing from adolescence to perfect manhood.

With females, on the other hand, the original flexibility of the cartilages of the larynx remain without much apparent alteration; thus they can sing in the same tones through life. Their voice remains always the same. No such physiological metamorphosis occurs in them, as in the boy, that alters the shape or cartilaginous character of the vocal box. For expansion or development of the larynx, its powers were very slowly maturing for thirteen or fourteen years, and when the voice changes, Nature announces, in that sudden evolution, an extraordinary physiological revolution in the boy's system. He is then a man. His beard grows, the muscles attain more volume, and all the powers of the body and mind are exalted.

This law perplexes physiologists. They have not successfully explained vital phenomena which still await elucidation.

Why some organs are active, and others quiescent for successive years, and then quickly burst into vigorous development, waits the patient researches of future philosophers.

In height, weight, and corporeal beauty, women differ from men essentially. In their moral constitution they also differ. Although neither so tall, so heavy, nor so strong, they are not without a commensurate compensation, always equal, and in many respects more interesting, according to the progress of a refined civilization.

### LAWS OF LIMITATION.

In comparing the physical structure, it must appear obvious to the most superficial observer, there are laws in force which regulate and determine animal growth. The elongation of the body of a man much beyond the stature of six feet, is a deviation from a normal standard in nature. An inch or two above or below the ordinary height excites no particular surprise, as a departure from the ordinary standard of humanity; but six feet and a-half or seven feet are anomalies, arresting our attention as abnormal, and, therefore, extraordinary. An experiment of an eccentric King of Prussia for rearing an army of giants, by compelling the tallest soldiers to marry the tallest women in the kingdom, exclusively, whether they were willing or not, is a matter of history, which proved singularly unsuccessful. The children of such parents, as often as otherwise, presented all those intermediate conditions between short and tall, characteristic of families in general. In a group of six or eight descendants from those unusually tall fathers and mothers, perhaps a majority of the sons were six feet. A few were even taller than the parents, while the remainder fell considerably below.

There are representatives of those Anaks in various parts of Prussia, at this time, but nowhere are there either families or communities which have perpetuated an unusual altitude. They degenerated to the original measure determined by a recognized law of limitation, and there men and women will remain.

There is a seeming predisposition in tall men to select short women for wives. It is an inborn, inexplicable fancy of exceedingly tall women to marry short husbands. It may not be either universal or imperative, but it is so frequently occurring as to have been noticed by philosophical writers, earnest interrogators of Nature into causes and the effects of causes. They think they perceive in this spirit of selection, otherwise defined to be an impulse of affection or preference, a law for equalizing the height of mankind. Were dwarfs to give preference to the marital companionship of dwarfs, and giants to giants, there would possibly be the two extremes—pigmies and Brobdingnags dividing the habitable portions of the earth between them, instead of races of rational beings controlled by a uniform law of limitation, standing upon the same plane, and averaging the same stature.

Very tall men, with remarkably tall wives, are met with everywhere, but they are exceptions, rather than illustrations of the law of development.

# CHAPTER III.

## CURIOSITIES OF ANATOMY.

Laws of Adaptation—Reference to Lactation—Pelvic Carpentry—Exposures
to Weather—Being too Delicate—Progress of Sentimentality, etc.

PECULIARITIES of the collar-bones, the width and depth of the pelvis, articulation of the thigh-bones and some other deviations in the female skeleton already adverted to, are quite sufficient for establishing the truth of one important proposition; viz., that they incontestably prove design, and, therefore, there was a designer. As we cannot add one inch to our stature, or make ourselves either handsome or ugly, we are at liberty, as free agents, to improve our condition. The form given us in the shape of those few bones, proves as clearly as grander exhibitions of Omnipotence, the controlling agency of a superior Being.

Let us analyze a little further those few specimens of design, to gather further insight into the object contemplated and the results to follow.

These long collar-bones are braces, keeping the shoulder-blades from being drawn too far forward by the pectoral muscles. Were they to encroach, it would be to the injury of the breasts, crowding them out of place, and thus interfering with the prescribed function of lactation.

A female breast offers a more inviting pillow for the infant's head than the hard flat chest of a man. A woman who has had no experience in the care of children always exhibits more tact and success in managing them to their satisfaction, than the most tender, sympathizing man. The softness of the investing tissues of the chest, the wider space for the infant's resting place, and the delicate cushioning of the ribs, forms an object contemplated from the beginning, and indicated beyond question by their precise position.

The out-spreading hip-bones, give a breadth to the female form, which is striking in any form or fashion of dress, compared with the hips of an adult man. The pelvis is built up of only three immensely large, irregularly-shapen pieces, constituting its walls. The key, or wedge-bone, under the name of *os sacrum*, is the base on which rests the spinal column. Its prolongation on a horizontal line in animals is the tail, but which in the human skeleton is formed of several distinct pieces, gently curved, so as to become, at the extreme tip, a flooring of the pelvis, for sustaining the viscera above.

This particular section of the frame of the female, abounding in curious manifestations of means to ends, is complicated with muscles and vessels, and, consequently, cannot readily be described in a way to have the mechanism understood, without drawings.

# CHAPTER IV.

## SOCIAL STATUS OF WOMEN.

CUSTOM sanctions the treatment of women as though they were unable to bear atmospheric exposures, or meet hardships of any kind with impunity. It is a mistake.

When their lives are cast in pleasant places, and they are sustained by a conscious independence of circumstances, which can only be realized in a state of Christian civilization, they then present themselves in the dignity of intellectual character. Uneducated, and simply occupying the position of a slave or an out-door laborer, they are adequate to the severest test of servile employment.

In refined communities, where contentment prevails, and where she is contemplated as a dependent appendage, rather than an efficient assistant, woman physically deteriorates. Kindness may degenerate into sickly sentimentality. Lamb-like and gentle, restless, irritable, and presumptuously exacting, are the poises that have much to do with the happiness or unhappiness of the sex.

Industry being honored as a virtue, idleness, consequently, tips the beam in an opposite direction. Being unemployed is no mark of a lady. Those who imagine it degrades them to be associated with pursuits indicative of labor, unfortunately for themselves lose what they most covet,—viz., the admiration of their friends.

In-door industry is, by general consent, commendable, and there it is supposed that woman is in her appropriate sphere. The cares devolving upon them, married or single, relieve them from those exposures which bronze the features, harden the hands, and destroy those traits of gracefulness which largely contribute to the withering of their charms. Caring for children, presiding as the spirit of order in the domestic circle, or giving the products of the field their preparations for the table, are not incompatible with elegance of manners, courtesy, and the handy disposition of the toilet.

When she steps beyond that assigned theatre for the exercise of her powers, whether improved by education or displayed in the rudeness of untutored abandonment, a woman is out of place.

Whether wise or foolish, learned or ignorant, poor or rich, beautiful or ugly, it is conceded by most men, not by reasoning but by intuition, that woman should be favored, and not subjected to the same discipline, in any department of industry, as themselves. On this sentiment civilization took its rise. To some extent, savages and barbarians concur with philosophers, that females cannot endure as much as men, because they have not the same hardy organization; so they alternately favor or oppress them, regarding them as servants, but not their equals or companions.

With savages, woman bears all the domestic burdens, suffers indignities patiently, and rears up children tenderly, protecting them with a mother's undying love, to be abused by them as soon as they have strength in their little arms to give them a blow.

A true history of the world is also a record of the wrongs of woman. Her happiness, her sorrows, her influence, and her misfortunes, are not estimated as they should be. She deserves heaven as a compensation for her bad treatment on earth.

### EQUALITY OF THE SEXES.

In theory it sounds well, when a political demagogue prates loudly before a multitude, of human rights and the equality of the sexes. A millstone hangs as heavily at the neck of a colored woman on a cotton plantation, as it would suspended from the neck of the orator's wife; but circumstances alter cases. After election, nothing more is heard of all being born free to pursue their way to happiness, till preparation for opening the polls comes around another year.

Unfortunately for the best-contrived plans for ameliorating social distinctions in this commercial age, it is necessary to stand on a pile of dollars in order to receive the same attentions accorded to those who actually possess them. Talent, education, or blood of martyrs in one's veins, are no recommendation to an acquaintance with property-owners, because revenues are the accredited touchstone to respectability.

Every city in Europe and America has its philosophers in rags, splendid women in poverty, the descendants of great families without a shilling. Who cares for them? Who invites them to dine when they entertain distinguished guests?

Nobody! No, they are not asked to take a seat in the broad aisle of a church erected by their ancestors! This is a text for reflection, but not a suitable subject for a sermon, it would so shock the sensibilities of devout hypocrites, who worship mammon under the mistaken idea of honoring the institutes of religion.

### AN INCONSISTENCY.

A glaring inconsistency in the present order of society is an unwillingness to allow females to sustain themselves by industrial pursuits which are claimed to be the legitimate avocation of men. By their exclusion, therefore, from enterprises perfectly within their sphere, many are unemployed, while another portion are cruelly overworked.

A million of women in the United States, and perhaps twice or thrice that number, contribute nothing to the common weal. Necessity makes no demand upon them, and consequently they are not only non-producers, but are sustained by the industry of others.

One class of unemployed females are denominated ladies, because they are above labor, and another stigmatized as vagrants, if scrutinized legally, on account of doing nothing.

No woman can be so far elevated by the adventitious circumstances of having an income that defrays her expenses, as to be exonerated from a moral obligation of doing as she would be done by, in her intercourse with those less fortunate than herself. Where that golden principle is lost sight of by man or woman, deterioration follows. A few are lulled on down and pampered on delicacies; others have measured out to them bitter draughts: vexations, disappointments, blighted expectations, thwarted aspirations succeed each other in rapid succession. Their pathway in life is through darkness and personal sufferings.

No wonder an expression of despair escapes the lips of those who feel themselves born to misfortune, in contrasting their condition with others, who never had an ungratified desire. They cannot see why they have been forced into existence to be miserable.

God in his wise purposes will clear away the clouds which make the course of life obscure to our limited mental vision. A law of compensation exists on the statutes of the Sovereign ruler of events, which will never be repealed while the pillars of justice sustain an edifice in heaven not made with hands.

Those difficult, social, and pecuniary problems are not for us
to solve. Why sinners are rich or saints poor, cannot be satisfactorily explained by human wisdom. Divine government is
impartial, since it rains upon the unjust as copiously as upon
the just.

## Might and Right.

Argue as we may to persuade the favorites of fortune to
divide their goods with the destitute, they will not do it. Inside passengers pity those exposed to the peltings of the storm
outside, but they do not voluntarily exchange places with them.
Nor do the poor, when unexpectedly put in possession of an
abundance, manifest a grain more of compassion than those
they before envied on account of their independence, or denounced for their cold-heartedness and want of sympathy.

## One of the Greatest Inventions.

Money was as potent when Abraham wandered with his
flocks as it is in the transactions of bankers in this year of
grace.

It was a great invention, and clothed with additional interest, when we reflect upon it, that whoever hit upon the idea
first, of having a piece of metal represent the value of a camel,
a horse, goods, chattels, or territory, thousands of times larger in
bulk, and then succeeded in making those to whom the scheme
was divulged agree to it, still more extraordinary.

Antiquarians cannot decide the epoch of its first appearance in trade. As far back, however, as sacred or profane history reaches, money was quite as potent as it now is. So precious was it very anciently, it was probably counterfeited,
which is inferred from a transaction mentioned in the book of

Genesis, in which the purchase amounted to four hundred shekels, "*current money of the merchant.*"

Civilization, from its phases in the orient to Anno Domini, 1873, has not much improved this universal representative of wealth.   It also represents power.   The mere belief that an individual has more of it than another, gains ascendency for him over those who were before his equals.

# CHAPTER V.

## EXTERIOR OF THE SEXES.

WHY the male lion has a shaggy mane, a larger body, or stronger claws than the lioness, is beyond our ken. Throughout the animal kingdom, with a few exceptions already partially recognized in a preceding chapter, males are larger and stronger than females of the same race, and far more beautiful.

Male birds, from the gaudy peacock to the ground sparrows, are magnificently ornamented with variegated plumage, difficult to imitate successfully by art. But, on the ascending scale, on reaching human beings, there is a reversal of the law. Woman's beauty transcends all other displays of beauty, while man is far less engaging in facial expression. His face inspires a different kind of surprise, admiration, or sentiments, but no sentiment of adoration.

Man's face is partially covered by a beard, if he is fully developed. His features are bolder, harder, and his build and movements are indicative of strength, vigor, and the wildest exhibitions of impetuosity. With massive limbs and regular deportment, he has no exterior beauty to be compared with the exterior of a beautiful woman. A handsome man is handsome by contrast, in possessing those harder, bolder, and rougher physical signs of attributes which animate him.

Why was a beard bestowed upon man? That question, many times answered, is still open for a more satisfactory explanation than has yet been given. If it serves as a sieve to prevent the inhalation of dust into the lungs, why not protect a woman in the same manner? She crosses the sandy deserts of Africa with her bearded nomadic associates, exposed to the same simooms. His cervical glands, *a priori*, require no more protection than those on her throat performing precisely the same office.

Alternations of heat and cold do not interfere with the functions of those salivary organs in her exposed neck, any oftener than when matted with a bushy beard. Nor do we admit the cogency of the argument, that a beard is a sign to signify the perfection of manhood. Those fair-skinned or dark tribes, or the red Indians of this Continent, are provided with no such appendage. The Caucasian has a beard. We shave it off daily, but Nature takes no hint—cares nothing about the inconvenience to which we are subjected in removing it with a dull razor: it continually grows. It was intended to subserve some useful purpose, but at present physiologists cannot agree what that is. Were non-bearded men mentally inferior, or those persons less muscular, a clue would be found to a solution of the question. There are as many Samsons without a beard, and bald, as there are with long locks and a disgusting sheet of tangled beard swaying over their linen bosoms.

Straggling hairs on the chin and the angles of the mouth on females are taken, on slender authority, however, as indications of a masculine character and sterility. A Spanish woman was extensively exhibited a few years since in all the principal cities, who had a prodigiously black bushy beard. She was the mother of three children, neither of whom appeared to have inherited a predisposition to its mother's anomalous appearance.

In her case, the development of a beard did not diminish a womanly expression of refinement and feminine excellence, nor did it interfere with any maternal relations. It was thick, glossy, long, which, with thick-set whiskers, would have been the delight of scores of beardless bucks who have vainly coaxed for a show on a smooth chin, through costly pots of perfumed bear's-grease.

When women have passed the age of maternity, it is quite common to be annoyed with straggling hairs shooting out on the upper lip, and about the lower edge of the chin ; but attempts at removing them by violence, the grip of tweezers, or jerks by the fingers, by creating a slight local inflammation, furnishes an extra determination of blood to the locality, that rather augments the crop. Depilatories are to be had in the shops which remove them without inflicting an injury to the complexion.

## FEMALE VOICE.

A particularly sonorous voice is ordinarily associated with a beard in men. The tone of the female voice is subject to none of the changes which the boy's larynx produces on his voice in passing through a pubert revolution of his system. The girl of the age of the boy is more mature, and shows her advance beyond him in the contour of her chest. Both remain physically stationary for many successive years. At forty-five or fifty, depending to some extent, perhaps, on climate, all other circumstances being equal, she passes through a change quite as curious and inexplicable as any phenomena which are stumbling-blocks in science.

With all her faculties in maturity, in health, in capacity for all the responsibilities belonging to her surroundings, nature is inflexible by declaring she shall no longer exercise the functions of a mother—she can no longer bear children.

On the other hand, man may possibly be a father at any period from youth to a full one hundred years, if reliance is to be placed in the statements of very high medical authority.

Some men's voices are not essentially altered in timbre at puberty. They are harsh, unmusical, or squeaky, which is attributable to an arrest of larynx development while other revolutionary changes are taking place in the natural order of events. An analogous transition in the vocal apparatus of fowls is noticeable. Young cocks make laudable efforts at crowing, which are ridiculous, compared with the full sonorous voice of a fully-grown chanticleer. Wild fowls exhibit no very noisy vocalizations like crowing. Theirs is a repetition of one or two notes or warbles. A sonorous voice is due, in part, to an evolution of sinuses or apartments in the bones of the cheeks and frontal bone, in which there are large chambers, bearing a certain proportion to the capacity of the box in which the vocal cords vibrate. In eunuchs, these sonorous rooms for the reverberation of sounds are hardly perceptible. There are none in children. The plates of bone begin to recede for the formation of sinuses at puberty. They are extensive in the skull of the lion, whose roar is a terrific sound in those dreary regions where he prowls a monarch over beasts.

## Their Ribs.

From immemorial time a vague impression has been entertained among those most susceptible in the way of marvels, of course the most ignorant, that men have not as many ribs on one side as the other; and the reason given for it is simply this, viz.; that Adam had one taken out for the manufacture of Eve. A very ridiculous notion, without a single fact to base it upon. Every well-formed man has precisely twelve ribs on each side, twenty-four in all. Seven are long, articulated to the breast-bone through the intervention of elastic cartilages. Five on either side are short, articulated posteriorly to the spine, but their front extremities float loosely in the fleshy walls of the abdomen.*

---

* A monomaniac in one of the Western States, in May, 1871, undertook to extract one of his own ribs, out of which it was his purpose to make a wife who should come up to his ideal standard of a proper companion for a bachelor of means!

This curious arrangement in the lower ribs allows for the enlargement of the stomach and bowels, and the flexion of the body forward.

All the ribs of serpents are free at their anterior extremity, and move like feet in crawling, each being acted upon by a complicated attachment of muscles. In consequence of their peculiar articulation to the backbone by a kind of rolling ball and socket joint, those hideous reptiles are enabled to swallow their prey in one piece, even when the mass has a greater diameter than their own body; the ribs, being pressed off either way, react back, as so many springs, to compress the contents of the stomach into the smallest dimensions as the process of digestion proceeds.

In number, situation, and use, the ribs are the same in both sexes. The muscular cordage embracing them is also the same, and they bear the same names.

Even admitting it to be literally true that a rib was taken from Adam, which we have no right to doubt, deformities, malformations, or defective developments, we have seen, are not transmissible. If they were, then there would be a space for a missing rib.

An excess of members is not unfrequent, but in a majority of instances, when there are supernumerary parts, as an extra finger, extra ears, supernumerary toes, etc., they invariably appear to have belonged to another being. In the commencement of uterine existence, there were two germs; the growth of one being arrested, while some fragments becoming attached to the other, in the progress of development, were nourished and became a part of the living child.

In every case, supernumerary appendages are considered as having been the property of the blighted twin.

## STRANGE FREAKS IN NATURE.

A man advanced in years, awhile since, exhibited himself extensively, who presented the strange anomaly of the lower limbs of an infant protruding from just below the pit of his stomach. To the spectator it had the appearance of a babe half hidden in his abdomen.

Originally there were twins. There was an arrest of development of one, from the hips upward. The other portion became attached to the other at the point of union described, and then there was a second interruption. The limbs had attained their present size, when all further growth was completely suspended. Had there been no causes operating to interfere with the uniform law of utero-gestation, there would have been a pair of twins of equal completeness in form and development.

The babes that recently died, born at the West, whose bodies were united in a way to appear as though lying on their backs, with their heads in opposite directions, are a further illustration of this melting of two beings into one.

Occasionally twins are born united firmly back to back. The Siamese twins are held together by a large ligamentous mass, the division of which might peril their lives, no surgeon being willing to sever the connection for fear of a hemorrhage from arteries they might not be able to control.

Where there are two heads with only one body, as seen in the colored sisters who have been through the States, they are two distinct persons. This is certain, because the two brains pursue different trains of thought, utter words, and constantly show in their mental manifestations they are distinct in soul, though nourished and sustained by one body. It is quite probable, however, that it will be discovered there are two spines, and two distinct spinal cords, hereafter.

An agent who appeared to have a pecuniary interest at stake in this double-headed girl, proposed insurance on her life at an office in New York. A question was at once mooted, whether there were two or only one individual to be examined. There were four lower limbs, but only one set of bowels, and, as it was thought, only one stomach. A paper was handed in from a medical gentleman of Boston, who gave it as his decided opinion there were two persons in the one!

In the course of these deliberations, we shall endeavor to show that defects are not propagated to the injury of a race. Individuals, but not families, are imperfect in form. Nature is conservative and corrects deviations, but never perpetuates them. Accidental circumstances modify conditions. Hence, the children of such deviations from a normal standard are not like their parents. One-arm children, children with only one leg, or those with extra limbs, are not, as a natural consequence, the offspring of parents thus defective or over-burdened with useless appendages.

### PELVIC CONSTRUCTION.

Notwithstanding the consideration that has been given in preceding pages to the pelvis, as a piece of mechanism, unrivalled, curious from the simplicity of its construction, and the many essential offices it sustains, it would be unpardonable to omit pointing out to parents, instructors, and those having charge of school-houses, seminaries, and institutions for the education of females, a danger that should be avoided, but which rarely receives any thought beyond the lecture-room of a medical college.

In a sitting posture, the weight of the body is transmitted

to the seat through the lower ends of two bones, having an irregular knob-shape, called *ossa ischia*.

If in early youth those three bones composing the pelvis, are forced out of place, or gradually distorted by pressure, it may not only produce a subsequent life of misery, but absolutely be a cause of a painful death to a woman.

As repeatedly asserted, the bones are slow of growth, and not completely ossified till near the twentieth year in females. A neglect to provide them with soft cushions or elastic coverings, instead of hard benches, hard chairs, or harder stools, while pursuing their studies, may produce such deviations in those bones as to be ever after beyond relief. A hard bone out of shape, or forced from the line it would have taken had it not been for habitual violence, cannot be pressed back to the position it should have to secure the benefits of a perfect organization.

No school for female children should be considered suitable for them, if the scats are not as generously supplied with cushions as the pews of a church.

The same danger does not threaten boys on board-benches. Their pelvic bones are set nearer together, are stouter, heavier, and the depth from the pubic brim to lower margin is shallower. In a word, on the perfect form of that bony basin depends the existence of the human race.

There is no parallelism between female savages and delicately nurtured young ladies, the pride and the glory of civilization. The latter cannot endure the privations nor sustain themselves under a tithe of those vicissitudes which are incidental to nomadic life. While civilization brings out the moral and intellectual faculties of an immortal soul, it carries in its train customs, habits, and tendencies which sometimes debilitate, undermine, or effectually destroy individual constitutions.

We cannot dwell on all the points that present themselves on reflecting upon what and how we are to act in regard to favoring the proper development of yonug females. They demand far more attention than they receive in the way of delicate attention. There is a public duty and obligation to be discharged, independently of parental solicitude. Providing them with soft seats in schools and seminaries is indispensable, and for the reasons here set forth.

# CHAPTER VI.

## IMPERFECT DEVELOPMENT OF WOMEN.

IN no country are there so many imperfectly developed females as in this, in proportion to the population. Nor are there more perfectly formed ones on the globe.

When a woman is defective in physical development, there is sometimes a corresponding imperfection of mind. Excessive nervous irritability, or any deviation from an uniform expression of that calm, consistent deportment which is a commanding element in the character of a lady, may be due to some derangements in her system.

It is proverbial that women of the Eastern States are spare, sharp-featured, and wear an anxious, restless expression. There are smiling faces, and fair ones too; but most of them exhibit an air of haste, nervous agitation on slight occasions, quite at variance with that gentleness of manner, sweetness, and affability, which, properly directed, wins more than a park of artillery could control.

Climate is chargeable with many influences which derange temperaments. Nevertheless, it is sadly to be lamented that, while some are constitutionally less attractive than others, it is their misfortune to make themselves unnecessarily repulsive. Assuming they have a presumptive right to do as they choose, and all men are bound in courtesy to bear and forbear under a galling fire from their batteries, such women are more dreaded than loved.

Women who are resolved upon driving, mistake their

mission. Weak men may be led by them; but it is a difficult
undertaking to drive those they may most desire to have at
their mercy.

No woman who has arrived at eighteen with a flat chest,
is harmoniously developed. Prominent signs of womanhood,
the absence of which are indications of a defect to be deplored,
not because she is less vivacious, less capable, or less able to
fill the rôle prescribed to the sex in the ordinary pursuits of
life, are very common with irritable temperaments. The
ingenuity of dressmakers and india-rubber manipulators is
consequently invoked.

There are young ladies, in the ratio, perhaps, of ten in
a hundred, in the Northern and Eastern States, on whom
there is no mammal elevation till they become mothers.
When that event occurs, there is an immediate deposition of
fat round the lactic ducts to protect the breast from injury
during lactation. At weaning, the adipose deposit is absorbed,
and the vessels, so carefully surrounded by elastic tissue against
the possible contingencies of contusions, while the fountain
was supplying the wants of a new being, shrink back to the
surface of the great pectoral muscle, hardly larger than fine
threads.

Fashionable interference with nature is the secret of this
anomalous condition. To an extent quite noticeable, the cut
and fit of garments suppress the mammal characteristic of
perfect womanhood.

It is a tacit acknowledgment in trade, that art takes the
place of nature in all cases where show answers all the
purposes of substance.

Artificial limbs, wigs, cambric breast-cups, basket-work
convexities, wooden calves, etc., which improve the appearance, are neither violations of statute or social law, and,

therefore, will not be abandoned while one sex has a desire to appear well made to the other.

So artistically are mammal appliances put in place, respiration produces all the movements as when the organs are in full maturity.

An imposition is practised both on old and young ladies of non-mammal development condition, that should be exposed. There is on sale in shops an ointment, exceedingly precious, according to the shameful misrepresentation of proprietors, for promoting the growth of the breasts.

Medications, either externally or internally, for that purpose, are positively useless. The swindle is enormously profitable, because no female, after wasting as many dollars as she has teeth, has the moral courage to denounce the fraud. It would be confessing her failure in the experiment. So the sale goes briskly on, and will till something new, represented more potent, with a sweeter odor, takes its place. Empty-headed bucks and beardless fops patronize whisker fertilizers in the same way, without ever having started three hairs where none were designed to grow.

### ANOMALIES.

When the hair bulbs are wanting, or are but imperfectly developed, which are hereditary conditions in some families, no medications are effectual in quickening them into activity. When they are imagined to have been serviceable in promoting a growth of hair, it is from friction in rubbing on the article, and not the preparation which produces the change.

Anomalous peculiarities show themselves from generation to generation in families. A predisposition to baldness is one; a beardless chin is another. But such departures are not uniform. Thus one son has a full beard and whiskers, while a brother is deficient in both. It is to be observed, in · regard to these deviations from a normal type, or inconstancy in external markings, that there are no departures or variations in organs essential to perfect nutrition.

Through the entire history of the Kendalls, as far as chronicles refer, a child is occasionally born with six toes, on one or both feet, or with an extra finger outside the small one, on one or both hands. But that by no means warrants a belief the Kendalls of England, or their relatives in America, are the lineal descendants of extinct Palestine giants, who were thus provided with additional toes and fingers. It is rather to be explained on the philosophical principle that has already been suggested, viz., that each and every supernumerary finger or toe is the remnant and only surviving one of a blighted twin, that would have been born had all its parts been symmetrically developed in time.

A female dwarf is often seen in New York, petitioning for charity, whose arms terminate at the elbows. There are no fore-arms. On the end of each stump are fleshy kernels, which may be properly considered rudimentary fingers. This is an instance of arrested development, and not to be confounded with cases of excess of members. Her lower limbs are perfect in shape, but not elongated, which indicates a second arrest of vital force at the period usually most active in children, when the shafts of their cylindrical bones are lengthened.

There is a much-respected member of the British House of Commons who never had arms or legs, nor are there any rudimentary prominences to lead to the supposition they ever had a germinal existence. Melancholy as this extraordinary form of defective external organization appears, he is a man distinguished for a brilliant and cultivated intellect. With singular adroitness, he writes with a pen in his mouth. That, too, shows to what vicarious uses muscles may be trained, and how the nerves, even those emanating from ganglionic centres, may conduct volitions or carry influences widely different from those assigned to them by the physiologists.

## ABNORMAL DEVIATIONS.

The subject is not yet exhausted. Some further observations on the fruitful topic of deviations are appropriate. A violation of a natural law does not abrogate it. It may be more logically expressed by repeating the words of another chapter, that a law of nature cannot be altered or abolished. In those singular deviations in animal forms from the true type, we see that a constant effort for a correction of the error or defect is apparent. Nature never relaxes or abandons the undertaking till the object is fully accomplished.

A calf with two heads, a pig with only one eye, a chicken with four legs, or a Nellis without arms, is a departure from a prescribed pattern. They are aberrations, and therefore not to be repeated by direct propagation. Whenever they happen, it is due to circumstances which we have not had the sagacity to detect by scientific researches.

Physical defects that incapacitate individuals from serving themselves according to the requirements of their nature, and for aiding and assisting their offspring till they are in a condition to take care of themselves, independently of the parents, are not represented in their progeny. Monsters are neither the fathers or mothers of monsters. Were it otherwise, confusion would follow, and no two animals would resemble each other in form, in character, or habits. The world would teem with frightful creatures, more hideous and terrible than the prolific imaginations of poets muster for their most daring contests with strange beings, created for special occasions.

## ORIGINAL FORMS PRESERVED.

By ingenious, persevering manipulations, flowers, fruits, and even animals may be produced wholly unlike those from which their origin was derived. But they cannot be kept at that point. A tendency to fall back to the form and condition of the original type cannot be effectually suppressed. A gardener's treatment, unrelaxed, furnishes the market with uncommonly large strawberries; but a relaxation of his attentions would be taken advantage of by vigilant nature, to turn them back to the size to which the law of limitation had assigned them.

Animals may be so amalgamated by interfering with the laws of reproduction, as to bring into being forms that indicate an origin from mixing races. They may not very accurately resemble either parent, and yet there are characteristic peculiarities which belong to both. Mules are neither horses nor asses. Without the beauty of the first, or insignificance of the latter, they are highly-prized hybrids, often taller than the horse, longer-lived than either of the parents, and with a hardier constitution, greater powers of endurance, and immunity from diseases to which both are incident. With such excellent properties, mules do not breed mules. Nature is consistent with herself in the enforcement of laws for the preservation of species.

## ORIGIN OF SPECIES.

We shall not meddle with the engrossing subject of evolution, the present plaything of scientists. Whether we are

degenerated monkeys or the children of Adam, is of no consequence in these investigations. It is enough that we are here;
but how or when the first human being assumed the prerogatives of a man, if the Mosaic cosmogony is ignored, cannot be
determined by quarrelling with theorists.

To fortify the position assumed, that nature does not allow
of the reproduction of defects, or rather deficiencies, of parts
essential for individual protection, further illustrations might
be collected quite as cogent as any already cited.

Of a large collection of remarkable examples, two more only
are introduced, not so much on account of their novelty as to
preserve a connecting fact, usually omitted, viz.; that persons
born without a full complement of limbs feel no deprivation on
that account, nor would they ever repine over the misfortune,
were they not commiserated and educated to a knowledge of
their condition.

A bank clerk resided in Boston, born with only one perfect
arm and hand. The other stopped short at the elbow. Exceedingly expert in handling bills at the counter, he could not
conceive of any use for another hand if he had had one.

Mr. Nellis, whose name was once familiar from Maine to
Georgia, was born without arms. Not the slightest rudimentary
elevation at the shoulders indicated a blighting of elementary
limbs. His skill in using scissors with his toes, writing legibly
and rapidly, drawing, handling a knife, firing at a mark with a
bow and arrow, was very surprising. He was a well-informed,
intelligent person, whose conversation and deportment were
those of a gentleman of refinement. Mr. Nellis frankly stated
that he could not realize that he was defective in any essential
particular, because he had no use for arms if he had them.

While waiting for a train at the western depot in Boston,
some years ago, a tall man came to the stove to warm himself,

whose hands were on his shoulders.  They were large, and the fingers long and bony, having the appearance of being used in laborious pursuits.  The arm-bones were there, no longer than at birth, but stout and strong.  The case is without a parallel in the writer's experience.  Had he been questioned, no doubt he would have said he experienced no particular inconvenience from the deformity, because he had not been deprived of any better arms.

#### ₁INDUCED MODIFICATIONS OF FORMS.

Mr. Charles Brown, a native of Waltham, Mass., died in 1871, whose right arm-bone, between the shoulder-joint and elbow, was absorbed completely, and carried out of the system. An injury, inflicted by a blow from the horn of an ox he was visiting in the stall, produced inflammation, which, without much pain, and certainly before there was apprehension of danger, resulted in that most extraordinary removal of a long cylindrical bone, without the escape of a single particle through an external aperture.  The brace being taken away which kept the muscles extended, they drew the elbow up to very near the shoulder, bulging out, of course, in shortening, by contraction, destroying the symmetry of the arm.  When his fingers grasped an object, or he lifted a laden basket, handled the reins of a harnessed horse, the arm was elongated to the original length. On letting go, the muscles would instantly contract like india-rubber straps.

With animals, when there are anomalies in respect to limbs, there is commonly an excess rather than a deficiency.  We have seen a dog without four legs which had acquired a method of going ahead with a degree of fleetness quite surprising.

It is possible to very materially abridge the growth of parts, to distort bones, and to promote or diminish vital force in the rearing of children. Civilization is imperfect when it conflicts with nature.

## THE DRESS OF LITTLE GIRLS.

They should never wear tight-fitting dresses over the chest. Entire and perfect freedom should invariably be given to that region. Any close contact of clothing over the pectoral muscles, or habitual compression, is an interference with a series of local changes, slowly progressing there, of incalculable importance in the economy of female life. Perfectly soft, pliable fabrics for their apparel need not be urged upon those who seek for knowledge in reference to a conscientious discharge of parental duties. For the ignorant, or those who care but little, for those who assert there is something more to learn, before we have exhausted the springs of thought, these comments are intended.

In the anatomical arrangements of the female chest, there is a congenital preparation for the development of organs at a proper time, the elements of which have been quiescent from early infancy. By and by compact cells are filled, and the mamma rise is organic completeness.

If, however, compression is maintained there, regularly and habitually, when an increased vital activity is preparing for the development of those organs, the contest will not be a protracted one between nature and opposition. Arterial energy will diminish under restraint, and the breasts will not rise, as they would have appeared, had no hindrance to the developing force been operating.

Even when there is a considerable adipose fulness unconnected with the mammary apparatus in its embryotic form, if close-fitting garments are habitually worn, the roundness and softness will be reduced, by absorption of the material deposited in the subcutaneous tissues.

Under the pretext for protecting the chest from cold, some mothers are despotically disposed to swathe their little daughters as closely as bandaged mummies. It is wholly wrong. Nothing should be allowed to interfere with the space between the shoulders in front to the tip of the breast-bone. Harnessing them in stays, corsets, or, indeed, any other contrivances of fashionable acceptance for improving the forms of young girls, are abominations.

By mismanagement, with the intention of improving upon nature, that they may have more attractions, and more arrows in their quivers when young ladies, mothers do them an irreparable injury.

### DEVELOPMENT OF VITAL FORCE.

Friction will partially raise the tone of vessels that minister to the mamma, but medications are inert and powerless in developing them where violence of dress prevented their growth at first.

Having shown the uselessness of lotions, unguents, electricity, or other trumpeted remedies for defective mammary development, and the grossness and unblushing impudence of impostors in that line of imposition, we proceed to another field where the harvest is large and the laborers few.

A withered or partially palsied limb may be improved by rubbing. The hand of a sound person is a thousand-fold better than a flesh-brush or hair-mitten. Friction accelerates the flow of blood where the circulation is sluggishly carried on, owing to the defective influx of nervous influence, which, together with warmth and the electrical current from the officiating operator, raises the tone of vitality in the member.

Women imperfectly developed are apt to be excitable, apprehensive, and wear the look of being cautiously watching for surprises. They are the women who are restless without cause, and unhappy in the midst of pleasant surroundings. They represent that class of ladies who are not treated as they consider they ought to be by their husbands. Conditions of the mind are recognized in which revolting crimes are perpetrated by women, not accounted for upon any well-established principles in mental philosophy, which, perhaps remotely, have a connection with some of those abnormal conditions of organs closely in sympathy with the brain, about which we shall know more when the progress of science has settled other questions respecting the phenomena of human life.

Mental feebleness may have been caused by a want of force from sources not precisely nervous centres. And the other extreme, of paroxysms of unbridled rage, arise from an excess of vitality, driven onward to engorge parts whose intimate relations to the encephalon are more direct than hitherto supposed.

## MEDICAL JURISPRUDENCE.

Medical jurisprudence is destined to undergo modifications, to keep pace with a more perfect knowledge of the brain, and especially the female brain, acted upon as it is by influences peculiar to themselves. When lawmakers have been educated to a comprehensive knowledge of the origin of nervous power, and particularly understand the phenomena of the passions, they may more reasonably account for many ungovernable freaks of an excited woman than are made easy of comprehension by writers on moral insanity.

In closing these monitory suggestions in reference to dressing little girls, it is hoped that no one may be so uncharitable as to consider it is impertinence to discuss a subject that actually has an important bearing on the physical well-being of female adults.

Physicians and tormented mothers know, by painful experience, of the origin of another misfortune; indeed a considerable one, too, which is, perhaps, caused by tight dressing, and certainly aggravated by it.

### SPECIAL GRIEVANCES.

Undeveloped nipples, far more common than supposed, are an interminable source of trouble, because an infant cannot apply its mouth for drawing milk. Artificial means for nourishing the child must necessarily be adopted, always to be deplored; and in the next place, the breast is injured by over-distension of the milk-ducts, or influenced by frequent applications of instruments for drawing off the secretion that would have been extracted with pleasurable suctions instead of painful inflictions, by the delicate lips of her darling.

That condition which gives employment to wet-nurses in the most fashionable circles, rarely occurs in the middle classes of society. Nature has her own way with children of the laboring classes. Little girls are not dressed and re-dressed in starched garments half a dozen times a day, to meet the requirements of dinner etiquette, the tea-table, the evening drawing-room, and various other specialties, to fit them for the positions they are presumed destined to sustain when of a proper age. Consequently they grow up in health, with the form they ought to have, and which the millionaire's daughters would have had, had they been simply let alone.

Who ever heard of a peasant mother requiring a wet-nurse? Where can a poor man's child be found brought up on a bottle, in consequence of the impossibility of taking its nourishment from the fountain prepared for it before birth, because the mother's nipples were prevented from developing by the indiscreetness of her mother?

In the fulness of our civilization, which is the triumph of reason over ignorance, we choke ourselves with tight cravats; ligate our limbs with straps, bracelets, or something equally objectionable, to check circulation; mount up on high heels, that force the feet out of the plane of comfort; wear patent leather, which prevents evaporation of moisture; cover our heads with airtight hats at the expense of our hair; sport with glasses that spoil our eyes; fill our stomachs with compositions productive of gastric derangements, and vainly seek relief from self-inflicted miseries that shorten life, in gorging with drugs that are worse than the diseases they were expected to remove!

# CHAPTER VII.

## The Dress of Women.

Small Waists—Sufferings from Fashion—Local Deformities—Compression of the Chest—Development of Consumption—Unheeded Advice—Form of Boys—How Female Dresses should be Worn.

Having explained, *in extenso,* the injurious effects resulting from improperly adjusted garments on female children when they are ·coming into womanhood, let us now investigate the positive character of modern female dress in respect to the production of disabilities traceable to that source in adult life.

Invention is, perhaps, exercised as actively in the production of new patterns, or modification of old ones, in the garments of women, as in any department of human industry. There is neither lull nor suspension in that most prolific field of restless variety. There is no stability in fashions. It is not required; since rest in that direction would be equivalent to a return to a system of simplicity and comfort identical with demi-civilization, if not barbarism.

Complete ease and freedom of the muscles seems never to have been contemplated in these ever-changing forms of their clothing; and the nearer they approach the borders of discomforture, without exactly killing themselves outright, the more agreeable, measured by a standard of the votaries of fashion.

## How the Chest is Injured.

It is singular that in the manifold styles of dress which succeed each other with the rapidity almost of barometrical variations of temperature, not one of them favors the freedom of the thorax or chest. That is the axle to which all pieces are attached, and the pivot around which they revolve, if at all.

## SMALL WAISTS.

A small waist is the first consideration. It is, therefore, the study of those who conceive they are too large just where there should be no interference with the respiratory apparatus, how to diminish their diameter. This desideratum has been the premature death of thousands upon thousands of the fairest and most promising young ladies, before they had time to learn the dangers they were inviting by following the example of those who teach by their practice that they prefer conformity to the requirements of a perverted taste, to exemption from the penalties of being out of shape, in the sense of those who exercise no judgment in regard to this important matter. The smaller the waist, therefore, the better, provided there is space enough preserved for descent of food to the stomach.

Stays are the instrumentalities for staying the development of the chest. Beginning early, the ribs are pinioned closely, and by unrelaxing ligation—the lacing being carried to the last endurable point without arresting respiration — their growth is arrested to an extent only familiarly known to anatomists. Their function is nearly destroyed, as they become anchylosed, or welded, where they were intended to have motion up and down, according to the inflation and collapse of the lungs. After being subjected to the torture of stays, for such it is, however eloquently those who have lived through the operation of having their chests kept down to the capacity of a child of twelve years, may argue to the contrary, breathing is with them an abridged function—or it may be arrested without hesitation. They have counteracted nature, and in various ways must, and do, suffer in consequence.

They even carry this violence to the chest still further, and interfere lamentably with the recti muscles in front of the abdomen, which reach from the pit of the stomach to the pubic arch. These are strong elastic straps for keeping the bowels in place and in contact. Thus, tight lacing forces the intestines out of place. One organ is driven too near another, and the stomach, instead of being pendulous, restrained by its own ligaments, is pressed down out of place, and that drags the spleen; while the free rise and fall of the diaphragm is limited, which strikes at life itself, because the lungs cannot be fully inflated when such displacements exist.

### DISPLACEMENT OF ORGANS.

After being worn till all these disturbances have become bearable, the distorted organs having been adjusted in new relations from which there was no escape, when a lady removes her stays she is very uncomfortable, because all those internal parts, acting in duresse, have a tendency to return into those natural relations from which they were forcibly driven.

Those abdominal muscles which keep the abdomen braced in symmetrical relations, entirely lose their contractible energy by being for a long while relieved from duty; and hence, in taking away the artificial support, the mass of viscera gravitates in a way to make a very undesirable abdominal protuberance in front. Hence, when broken into stays, the harness cannot be dispensed with without discomforture.

Chambermaids imitate their mistresses, as far as their circumstances allow, in self-imposed misery. Fashions and customs are infectious. When endemic, they have a regular run. Females, therefore, in the constitution of society, suffer more than men by the mutations of fashions. The latter make themselves ridiculous by the cut of their coats, the shape of their hats, or the show of toggery on their watch-chains; but they are too much afraid of dying before their time comes, to kill themselves with stays, although a few brainless fops make themselves extra-ridiculous by wearing them.

## WHO TO CONSULT.

If it is desirable for women to have convincing proof of the injury they voluntarily inflict upon themselves, that they may imagine themselves more attractive in the estimation of others, let them consult medical authorities. They will there have the collected opinions of professional men, who can have no motive for misrepresentations, that the sacrifice of women through the vice of dress, and destruction of infantile life, through malformations, displacements, and special maladies induced from the wearing of stays, is a melancholy comment on one of the demands of modern civilization. Ladies thus deformed, and in a part of the body, too, which prevents the respiratory organs and the heart from carrying on processes of importance to the vital status of the individual, look with disgust upon the little feet of a Chinese belle, kept down to the size of an adult great toe by bandages. They are cruelly served—not voluntarily. It is no self-inflicted torture; they uncomplainingly submit to make themselves more saleable, but it is forced upon them by ambitious parents, that they may bring a remunerating price for the trouble of rearing them. Of course, with such feet, they cannot walk with steadiness or facility. They must have support by a fan against a wall, a parasol, or the occasional touch of some solid resisting body, or fall to the floor.  And this is beautiful!  Nor is it a whit more absurd than disfiguring the chest, not allowing it to expand, nor half so injurious.  There are neither lungs or a pulsating heart in the feet, but there are both in the pleural cavities.

### MATERNAL INTEREST IN DAUGHTERS.

Maternal solicitude for the position and happiness of a daughter is manifested very differently in this country and China. There, no mother in whose bosom there is a grain of motherly affection, would be so lost to a sense of duty as to let her loved Ky-yan-ste shoot up to the stature of herself with feet as large as a Christian's.  No, indeed, that would be barbarous beyond forgiveness.

Compressing the waist with stays has precisely the same effect on the carpentry of the bones, that bandaging the feet produces.  When the violence is completed, the first cannot move comfortably without her stays, nor the latter hobble through a room without having her ankles secured by many yards of firm, strong, inelastic bandages, which, for show, are made of richly-colored ribbons.  Thoracic compression alters the figure of the lungs.  The chest is naturally broad at the base, becoming narrower at the top—a cone-shaped structure. Women make it narrow where it should be broad, and broader at the apex, where it was originally narrow.

### INFLATION OF AIR-CELLS.

The lowest air-cells of the lobes cannot expand when air is inhaled, while those in the upper region of the lungs are preternaturally put upon the stretch, in order to provide surface for the creation of blood. One end of each lung is compelled to do more towards maintaining life than it was organized to do, while the lower part is prevented from giving much more than a feeble degree of assistance.

Favored, as many robust women are, with a fine organization in other respects, they can live out a long life in comparative health and comfort; but they are few compared to the vast number who fall short and die before they have attained all they might have had on earth.

The first or topmost rib on either side, just under the collar-bone, is short, thin, and sharp on its inner curvature. It has no motion, being a brace between the dorsal column and the breast-bone. It is immovable for the purpose of protecting large arteries and veins belonging to the arms on either side of the neck. Such is the construction within the horizontal arch of that rib, the upper portion of the lungs rise up through the space above the level of the bone. In cases where the chest has been manipulated till the lungs cannot expand downwards, they are forced up above that rib. Rising and falling above and below that rib-level, the lobe chafes and frets against the resisting curvature. It is inflamed at last, and the organ becomes diseased. If that chafing is not relieved, but in each respiration the serous covering of the lung is irritated continually, the inflammation is apt to extend quite into the body of the organ, increased and intensified by exciting emotions, laborious pursuits, or unfavorable exposures. Finally, the mucous lining of the air-cells within the lung sympathizes and becomes inflamed also.

## COMMENCEMENT OF CONSUMPTION.

In this condition we may trace the commencement of pulmonary consumption. It would be denominated sporadic, and widely different from pulmonary disease by inheritance. But the possibility of deranging the function of the lungs by simply distorting the chest, cannot be doubted, nor would any anatomist presume to say such treatment does not do violence to those much-abused, delicately-constructed organs. Being forced from their normal place in the pleural cavities, is dangerous in the extreme.

Consumption is not only developed by tight lacing, but a multiplication of cases, where the original conformation of the individual was favorable for a comparatively long life, is beyond question. Medications cannot stay the onward march of disorganization, when ulcerations eat the tissues. Once destroyed, they can never be reproduced. Therefore, if prevention is better than cure, less expensive and always more agreeable, why not profit by these suggestions?

No compression of the base of the chests of men being induced by tight dressing, a chafing of the upper surface of the lungs rarely occurs with them. If, by constant effort to distend the lungs, the lobes extend where there is the least resistance, the tissues covering the space between the inner curve of the superior rib and cervical vertebræ gradually relax, and are convexed upwardly at each breath. This, therefore, explains the mechanical results of thoracic compression, and women, as a matter of course, are the most frequent subjects of a diseased condition of the lungs from that cause.

## UNHEEDED ADVICE.

In a blaze of hygienic light, admonitions of the medical profession are unheeded, and death and stays act in unison, decimating the fairest flowers of intellectual womanhood. A warning voice is lost in the distance when it refers to this subject.

Not one mother in a thousand doubts the truth of what physicians proclaim in respect to this painful invasion of the chest, yet she continues the practice.

Great men, giants in any department of busy life—those who make the world conscious of their influence—those who quicken thought or revolutionize public sentiment, and leave the impress of their genius in the history of the age in which they flourished, were not the sons of gaunt mothers whose waists resembled the middle of an hour-glass.

### TRANSMISSION OF DEFECTS.

Mothers certainly transmit their own physical, if not their moral and other qualities to their children. A feeble organization is perpetuated through successive generations, terminating at last in the extinction of a family, unless there is a revivification of vital force by an intermixture of a healthy stock.

Intermarrying among relations, with a view to a selfish purpose of keeping estates always within the same control, or from a spirit of pride that looks with contempt on alliances with other blood as contamination, cannot be sustained. There must be crossings, and an infusion of new elements. Utter extinction of a family may safely be predicted that tolerates no affinity with other blood.

Nature asserts the law, and, if not respected, a race cannot conceal its deterioration. A feeble intellect, supported by an imperfectly developed body, is a notification of a sovereign decree—the disappearance of a family—only to be saved by the formation of new relations with those who have vitality if they have not property.

### PREDISPOSITION TO MALADIES.

Competent medical authority has decided that a predisposition to certain maladies are transmissible from parents to children. Seeds of disease may remain quiescent many years, and then suddenly burst out into destructive activity. Changes of weather, variations of temperature, when an individual in whom they may exist is exposed, together with the peculiar susceptibility of such persons, produce slight inflammatory turgescence of the mucous membrane of the throat, which, creeping down to the interior of the air-cells of the lungs, assumes a very grave aspect.

The next phase in the progress of incipient pulmonary derangement is a cough. Purulent matter is excreted over the bronchial mucous lining of the air-tubes, to defend them from irritation from the direct contact of air on the inflamed mucous membrane. Violence in the attempt to raise that matter, which, of itself, is another source of aggravation, from its weight,—the thin partitions of the cells are often ruptured by spasmodic paroxysms of coughing. If not removed, the accumulation, remaining in a mass, ulcerates the membranes, and pus gravitates downwards. Abscesses are formed. Thus the integrity of the whole lobe is involved.

Emaciations, in consequence of organic derangement and imperfect oxygenation of the blood, is the result. Debility marks the onward destructive progress of ulceration. Neither tonics, the modification of diet, or a change of climate, can arrest the further destruction that must inevitably terminate in death, when the mechanism by which respiration is conducted is destroyed.

This is a mechanical delineation of the phenomena of

induced pulmonary disease, by violations of the laws of health. Dress, where it interferes with a perfect expansion of the lungs, certainly tends to the shortening of life.

## Not Curable.

Pulmonary consumption, in the form here described, cannot be cured; nor can it be much relieved. How absurd then, on the face of it, to fill the stomach with drugs, with an expectation of regenerating parts that have been completely destroyed. Nostrum-venders thrive by the sale of consumption-remedies, but they are the only persons benefited by their falsely-represented panaceas.

## Further Interviewing of Stays.

Women are not expected to lay them aside. While it is universally admitted by them that their taste is superior to nature, stay and corset-making will be a profitable branch of manufacturing business in coming years.

Why do not boys require such appliances? Without them, left to themselves, they grow up with full, rounded chests, and their proportions are admirable. Rare examples of feminine, beardless exquisites in stays are known at fashionable places, the straws on the ripples of society, but they are invariably regarded as brainless butterflies who are neither men in character nor women in form.

Criticisms on female dress are not the outpourings of an envious spirit, when they emanate from professional writers. Life is a boon so precious, they fain would persuade women to preserve it, and not sacrifice it to the caprice of fashion.

## How their Dresses Should be Worn.

Were all their large garments suspended from the shoulders, the consequences resulting from confining them round the waist with the grip of a boa constrictor would be obviated.

In addition to the close-lacing of stays, each lower garment is bound tightly on, over the same region, to keep them up. This is all wrong. Closely-pinned waists of petticoats, bands, belts, and buckled ribbons, girdles, or straps, positively stand with firm resistance to the development of the base of the chest. So it is perfectly clear without a labored dissertation, that the mischief habitually practised to the positive injury of the whole internal economy of the female body, might be avoided by simply suspending garments from the shoulders.

Very young children are thus dressed, a mode only to be abandoned before the bones of the chest begin to ossify at their distal extremities. Moral and mental circumstances in a little girl's every-day life are overlooked, comparatively, in the effort to improve their forms.

Of the amount of disturbance produced in the basin of the pelvis by constantly tying on garments, a detailed description is given in the lecture-room where diseases of women are explained. It is difficult to popularize the subject, and that is one of the reasons no more progress has been made in revolutionizing their costume. If a cord were daily wound around the body just above the hips, the bowels would be forced downwards, interfering with another set of organs. That is the true cause of a painful catalogue of maladies to which women are incident. Displacements cannot be inflicted without suffering and real danger. Multitudes of females reach an advanced age who have survived the misfortunes entailed upon those who possessed less tenacity of life, from being subjected to the ligating discipline of the waist. But that is no valid reason for continuing a practice so destructive in its tendencies.

## Proof by Analogy.

Some men escape injury in severe battles, where the ground is strewn with the dead and the dying. Is that a proof that others might escape also, exposed to showers of flying balls? Where one woman, apparently, has had no inconvenience from a diminished waist, more than one hundred have died.

The weight of heavy clothing suspended from the shoulders is not as burdensome as when suspended from above the hips. Still, with that fact before them, ladies have made no alteration in their mode of dressing. It is a favorite way of demonstrating the looseness of their garments about the waist, that their fingers can be pushed under their belt.

That is quite possible, but the extreme ligation is in the girded skirts that are worn, pinned, or buttoned as closely as they can be drawn.

Brigades of physicians thrive professionally, because women persist in making themselves sick. Specialists find their complaints a profitable field for culture. Female doctors, too, have not been unmindful of the advantage they possess in gaining the confidence of their own sex, by turning their folly to good pecuniary account.

Oriental females keep their garments in place by a scarf or shawl, according to their means. Their trowsers, immensely large, soft and pliable, are easy for the limbs, and graceful in appearance. They, however, are by no means exempt from contingencies which belong to girding the waist. Although pretty severely ligated, they are not injured in the same way that civilized women suffer. If they were as energetically industrious, they would be equally exposed. Their habitual indolence, especially the higher classes of ladies, the stars of the harem, is favorable for them. They have no chairs, but recline on elastic cushions. Were they obliged to exert themselves in lifting or carrying heavy children in their arms, they could not escape those mechanical displacements which are intimated, without being specifically described in these observations.

With them, their scarfs are not quite as terrible as stays. Instead of compressing the base of the chest, Turkish ladies make the ligation lower. They spare the lungs, but in stooping or rising suddenly, they are frequently ruptured. The bowels are forced to a point of least resistance,—the groins, where hernial protrusions are common.

## HERNIAL PROTRUSIONS, HOW PRODUCED.

Greek women have more freedom ; and engaging in domestic pursuits of all kinds, in consequence of keeping their clothing together, precisely as their Turkish sisters do, they are extensively and badly ruptured. Perhaps no country in the world furnishes an equal number of ruptured women.

Women are subject to indispositions peculiar to their organizations, which may be made worse by neglect, or perpetuated by continued violence, however gradually inflicted. On the whole, leaving Nature to herself, the sexes possess equal advantages for health and longevity.

### Simplicity in Dress Security for Health.

Permit little girls to pass their youth untrammelled by garments that would either compress, or in the least degree interfere with the chest or abdomen.

Distortions of the pelvis would be avoided by providing suitable seats at school, and also at home. No bone can be pressed out of line without interfering with some function that sooner or later may be a source of suffering or sickness.

While learning to write, their positions should be frequently varied. If they habitually sit in the same place, taking the same posture, there is danger of swaying one shoulder or warping it to one side. Young girls are more prone to have their shoulders distorted than boys. The latter are nervously using all their muscles, especially those of the arms, which secure symmetry to their shoulders. Girls are restrained from playing ball, climbing trees, or engaging in exercises that force the muscles of the spine to extra action. If girls are left too long at the desk, one set of muscles relax, while the other set are kept too long contracted, inducing weariness. Curvatures of the spine have their origin in not sufficiently varying the postures they fall into by occupying the same seat, the same desk, or receiving light from the same direction always.

Where a scrofulous habit exists, there should be even greater caution in varying the position often. Narrow chests, a breastbone pressed inwardly at its lower end—two sad conditions—may be avoided by the simple process of having the books on a high desk, which would compel the pupil to sit up straight.

With these statements and recapitulations of what parents and instructors should do to secure the health, vigor, and beauty of young girls, it is not pretended that perfect success will crown their efforts. Some of the most faultless in form die prematurely; but that they are wronged out of vitality they might have had, treated as boys are, in respect to clothing and out-door exercise, is mournfully true and lamentable.*

---

* In those parts of France in which stays have been laid aside as injurious, it is stated the mortality of females has decreased eighteen and a-half per cent. According to the same authority, chignons increased cerebral fevers seventy-two per cent.

# CHAPTER VIII.

## EXERCISE OF WOMEN.

FEET were intended for use, yet there are women quite unwilling to exercise them in any other way than dancing. Some
scarcely feel able to walk from a dressing-room to a dinner-
table after completing an elaborate toilet. Elegant idleness
cannot be persuaded that it is not vulgar to move about on
one's feet. Airing in a carriage is genteel and without
fatigue.

Anybody can walk who is not a cripple, but all cannot ride.
It is charming to take a pleasant drive, provided the weather is
perfectly agreeable. Greeting choice friends from the windows
of a splendid coach, in passing, is infinitely superior to plodding
along on foot at the risk of rude contact with disagreeable
people ignorant of the rules of good breeding.

An apprehension of damp feet by touching mother earth, is
a common excuse for not promenading like those who never
owned an equipage. The susceptibility to cold is quite surprising with some ladies who could once trip through the wet grass
when they resided in their country homes with impunity.
Moonbeams become too ponderous for their fragile nerves since
coming to the city and into the magic circle of fashionable
exactions.

There are occasions, notwithstanding such acquired delicacy
as passes for an unequivocal sign of social elevation, when even
such zephyr-like humanity rises in the dignity of heroic resolution, to mingle with the world in crowded assemblies, waltz all

night for charitable purposes, retiring at daylight the following morning, satisfied with themselves in having discharged a religious duty.

If it is too fatiguing to trudge on foot like servants, how much more to ascend long flights of stairs, unless they lead to the exquisitely furnished apartments of a friend. In their own dwellings they are not unfrequently carried in a chair or borne in the coachman's brawny arms from the doorstep to a carriage.

## EXTREME DELICACY.

It kills some ladies, in court language, to exercise in any ordinary manner. This is a common complaint of very sensitive beings who were once chambermaids or milliners. To appear perfectly well is to acknowledge themselves rather plebeian. In their early days, glowing with freshness, vigor, and the best elements of a sound constitution, it was the good fortune of many who now converse most about remedies to have captivated a prosperous groceryman, a thrifty tailor, or the rich son of a retired leather-dealer, who was accepted as a lesser evil than remaining at service. Exchanging a cot in the garret to become mistress of an elegant establishment on an avenue, is not to be despised. Their husbands pursue the tenor of their ways, multiplying goods and chattels, and becoming millionaires, while their wives develop into model patients, patrons of music, the drama, art, select dinners, the opera, and tract-distributions to the poor.

Before marriage thus advantageously secured, every close observer has known spirited young wives who could once run from the basement to the skylight without complaining. Now cologne out of a phial would not revive their exhausted spirits. A few years of technical luxury, surrounded and enveloped in comforts and elegancies to which they were unaccustomed in the elastic days of youth, they decline to an abyss of chronic indolence.

## MUST EXERCISE FREELY.

The less we use ourselves, the more rapidly we deteriorate. When muscles remain inactive, they lose their tonicity. They cannot be strengthened by taking drugs, but by proper exercise. Pedestrians derive advantages from facing the breezes, and communing with nature in the open highway, which the occupant of a carriage does not receive so advantageously. Her locomotive cordage is at rest while riding. The walker puts all the contracted fibres of his body in motion at the same moment, and, therefore, every organ feels the impulse, and is benefited accordingly, because there is an increased activity in the circulation and the secretions and exhalant vessels.

No form of exercise has been pursued which is productive of health-giving vigor, to be compared with habitual promenading on foot, regardless of weather or season.

If men delight and enjoy pleasant walks, why should not women? Alternately balancing the weight of the body on one foot and then on the other, brings every muscle to its full bearing. Each one of them has an antagonist, and thus tension and relaxation create a demand for nutrition, proportioned to the force they may be called upon to exert. An appetite is created to meet the wants of each and every tissue; and in providing for a hungry stomach, we simply feed each one of those muscular threads which assisted us in stepping off briskly.

Without appetite, strength fails, temperature diminishes—the extremities being cold—and direct debility is the next condition. Every limb, or section of one, may have its form

increased simply by exercising it.    Insufficient food reduces
vital force.*

Bearing burdens, hauling ropes, working at a pump-handle,
lifting kettles from a range, swinging a broom, etc., gives the
female cook beautifully rounded arms, the envy of her mistress, whose bony apologies for arms cannot be made attractive,
even encased in diamond bracelets.    Dancing develops the
lower limbs.    Riding on horseback brings out the full proportions of the chest and abdomen, but does not round up the
muscles of the legs like walking.    Ladies do not reap as much
benefit from that exercise as men; because only one limb has
opportunity for bracing, while the former press equally on
the stirrup with both feet.

Next to walking, a bracing morning-ride on horseback is
incomparably superior to an airing in a carriage.    Efforts are
unconsciously made on the saddle in maintaining a perpendicular position.    That is what calls out an extra effort of the
muscles, and hence they increase in size and power.    When
a lady drives out for the purpose of refreshing her debilitated
system, simply inhaling the fresh air does not accomplish for
her all that an uncontaminated atmosphere certainly would do,
were her muscles set in active motion at the same time.

As boat-rowing is wonderfully conducive to a broad,
rounded chest, we are surprised that it has not been urged
upon narrow-chested, feeble, consumptively-inclined young
ladies.    They would realize all the sanitary advantages from
an elegant and extremely popular gymnastic exercise, that

---

* Four of the wealthiest gentlemen in the city of New York, distinguished for their millions, dined together the last Sabbath of June, 1871.
They were famishing for want of appetite.    The rich viands were scarcely
tasted.    If each lived on sixpence a day, and earned it by labor, they would
not have complained of want of appetite.

those do who figure in clubs and rowing-matches. They present the finest-formed chests and the best breathing apparatus of any class of men.

A hint might be taken from the pursuits of professional bargemen. They have prodigiously large, fully-developed chests. Diseased lungs in their calling must be rare. With these views, the result of carefully surveying the tendency to invigorate the pectoral muscles and expand the thorax by handling oars, we strongly recommend boating for ladies of the description referred to in these observations on exercise. They might count upon having splendidly-rounded arms by that graceful amusement, and improved chests, if they would be sure to remove their stays before seating themselves at the rowlocks.

A side-saddle is very well, as far as it goes; but inferior to the man-saddle, inasmuch as the bracing is made exclusively by one foot, as already mentioned.

## NUTRITION.

Nutrition of the body is a very interesting subject, not generally understood, although a very frequent topic of conversation among those knowing the least about it. How few comprehend the phenomena of digestion. When food falls into the stomach, it is lost sight of, in the ordinary way of speaking. At that point a series of vital activities and changes commences, that have given rise to researches of peculiar interest.

While an animal is growing, it is taken for granted that food furnishes materials for completing that process. When full proportions are attained, the body is apparently stationary; but, by eating and drinking, materials are furnished for keeping it in repair. A waste all the while is going on. If that daily wear and tear were not met by a new supply, there would be immediate loss of weight and immediate debility.

Now comes into view the economy of nature, by which appropriate elements are elaborated from food in that membraneous bag—the stomach—which are floated along in tubes to places where new matter is required to take the place of old substance which has just been removed.

Arteries may be compared to canals, through the aid of which freighted boats carry every imaginable product of the country for meeting the necessities of the people.

Blood runs through these vessels, in which there is held in solution whatever is required—such as lime, glue, phosphorus, etc., too numerous to mention—which is carried to the remotest fibre, where each takes up what it needs, and no more; and whatever remains, after being thus selected from, passes on to other stations, where freight is discharged, according to the demands of the body.

The mechanical part of digestion is simply this. After being reduced to a greyish pulp in the stomach, by being mixed with a variety of products which have their origin in glands, food gradually enters the intestinal canal, a thin, strong, curiously constructed tube, about six times the length of the individual. In childhood it is nearly eight times the length of the body.

### LACTEAL VESSELS.

From the descending mass of food urged through the intestinal tube by its contractions from above, a milky fluid is formed called chyle. On the inner surface of the long tube are millions of minute openings of hair-like tubes which terminate in fleshy masses of different sizes, lying between the duplication of mesentery. Those little orifices suck up the chyle as it passes by, and convey it to the mesenteric glands. It remains in them but a short time, when it goes out through another set of minute tubes on the opposite side of the gland, to be conveyed to a small white tube lying in contact with the back-bone, known as the *thoracic duct*. In its exit from the gland, probably something is added, or some chemical alteration takes place that improves its quality.

The thoracic duct ascends by the side of the vertebræ, not much larger than a wheat-straw, till it reaches the root of the neck, where it curves and enters the jugular vein of the left side.

### How the Blood is Produced from Food.

At the angle, the white fluid produced in the bowels, the essence of food, as it were, mingles with venous blood. The current of blood and chyle mingling runs across the top of the chest just back of the breast-bone, and empties into the right auricle or upper chamber of the heart.

As soon as that apartment is full, the walls contract and force the contents through a round opening into the next cavity, the ventricle, which contracts and drives the fluid onward through the pulmonary artery into the lungs. That great vessel subdivides in the substance of the lungs, infinitely, into fine branching vessels, where each air-cell receives a twig that spreads around it like net-work.

Air is next inhaled, inflating those cells, and in the act of inflation, the oxygen of the atmosphere comes in contact with the newly arrived fluid, spread like a film around the cell. At the same moment, carbonic acid is thrown off. The imbibation of oxygen changes the mixture of old and new blood— which arrived together, as described in the jugular, into a rich

scarlet color. It is then a vitalized fluid, arterial blood, and
ready for general distribution by the contractile energy of the
left side of the heart.

Effete matter, that which remained in the system till it had
imparted all its serviceable properties, is evacuated. Thus an
explanation of the reason for eating and drinking is made plain
enough for the comprehension of a child.

## OUT-DOOR EXERCISES.

Exercise accompanied by pleasurable emotions, as from the
view of verdant fields, mountain scenery, flowers, or refined
social intercourse, is eminently calculated to sustain and im-
·prove our health. It should be encouraged by those having
the care of children. Public teachers should give it their
approval. In all institutions, educational especially, frequent
opportunities should be given pupils for free out-door contact
with the air, regardless of the season. Air was designed for
breathing. Those who have the privilege of being exposed to
it most, will appreciate its sanitary value.

Laborers have a compensation for their toil beyond a pay-
ment in money, in the sound condition of their bodies. They
are not always under the doctor's care. They have no fear of
an east wind, the dampness of a napkin, a hard-boiled egg.
They neither have dyspepsia or go to the White Sulphur Springs
on account of ailments generated by idleness. Women above
industry, gently driven in a close coach, lest a ray of light
should imprint a bronze hue on their pallid cheeks, envy the market woman, strong, hearty, and well, unconcerned about the
shade of silk, or the lace trimmings to be worn at the next opera.
Being used and not used, are very different conditions. Not
only health, but even the length of the thread of life are deter-

mined by the use or the neglect of our various powers. Pursuits which put the long muscles, as those of the back, chest, abdomen, and extremities into frequent action, are most conducive to continued good health.

Having particularized the benefits to be realized from horseback exercise, it will be found that those on foot gain more than riders. They are longer-lived, and are freer from attacks of disease, either acute or chronic.

Peasants in Europe, and females of the humble orders in Oriental countries, who carry heavy jars of water on their heads, make no complaint. Each and every muscle is brought into a taste of tension in the act of balancing burdens thus transported. They are exempt from spinal difficulties, being subject neither to dropsical effusions, spinal irritation, incurvations or curies of their bones.

### REMEDYING DISTORTIONS.

An orthopedic institution, which copies the Nilotic water-girls, requiring fragile female patients to support weights in the same manner, instead of requiring them to pass hours on an inclined plane, and the remainder of the day to be imprisoned in stiff, unyielding apparatus, would succeed far more satisfactorily than in the old way of going counter to the best indications of Nature. A weight on the head would immediately call into play the dorsal muscles, which would increase in volume and strength with repetitions. Strapping frail, slender, imperfectly-developed girls, as commonly practised, to boards, a hard bed, or lacing them in metallic corslets, with an expectation that a distortion is to be overcome by it, is entirely wrong. Gradually bringing into use neglected apparatus, as muscles of the back, chest, arms, etc., and with appropriate attention to diet, relief may be reasonably expected. Tonics will not give the wished-for relief, unaccompanied by exercise.

Young ladies of a lymphatic temperament, not disposed to exert themselves beyond what may be perfectly agreeable, who delight in lounging away the precious hours of opening life on elastic couches, or languishingly reclining in a luxurious coach, for an occasional airing, when the weather is unexceptionably fine, receive but little advantage from scientific treatment, when distorted, simply on account of the extreme tenderness with which institutions treat them.

Scrofulous, sallow, indolent, lachrymose, sentimental ladies, whose circumstances are ample enough to warrant them in gently descending to the grave in all the pomp and circumstance of fashion, would not submit to such manipulations as might turn the shadow a few degrees back on the dial of life.

## Social Phases.

Condition modifies circumstances. Some are unhappy because they cannot compass unreasonable projects; and others complain of being wretched on account of neglected claims to social position. A disgust of life is not an unfrequent apology for suicidal acts, which are charitably imputed to derangement of mind consequent upon ill-health. There may be a form of mental depression that so lowers the vital mercury as to make it appear easier to die than live in neglect or hopeless uncertainty of ever being appreciated.

There is another order of female despondents who are socially miserable by mistake, entertaining an idea they have not all an ungrateful world ought to give them, while they are revelling in the midst of phantoms and vanities. In a moment of desperation they swallow a dose of opium and slumber into eternity. This is a woman's way in distraction. Men blow out their brains with a revolver, or with a razor tap their jugulars.

Ladies in poor health, who cannot be miraculously relieved —the broken-hearted from unrequited love, victims of dissipation and the ignorant, who conceive themselves of more importance than others admit, those who are always trying new remedies from irresponsible sources, certified to by persons whose word is worth no more than their bond, those who consult quacks, have the blue devils, and refuse to be comforted, and require watching, there being a suicidal tendency—would each and all of them receive permanent relief from regular employment, coarse nutritious food, and daily walks that would invite sleep from fatigue, instead of taking medicine or consulting clairvoyants.

## WOMEN OF ENERGY.

Hardy, resolute, energetic women, who rarely ride, require no medical assistance, mineral water, or soothing compositions. It is so common and perfectly genteel to be most of the time an invalid, that it operates very unfavorably for the prospects of those who imagine it gives them an interesting appearance in the estimation of sensible men. They are unwilling to open a private hospital in entering upon the responsibilities of matrimony.

Robust, clear-complexioned women are not usually natives of cities. Those who have the true elements of that kind of womanhood which will best sustain them in city life, are transferred from the country. They bring with them a stock of vitality which resists the effects of a vitiated atmosphere and the debilitations of luxury, rather longer than those "to the manor born."

But warm apartments, coal fires, gas lights, late hours, rich food, strong coffee, and the pride of wealth, wear upon them after awhile. Women in health are the hope of a nation. Men

who excercise a controlling influence—the master spirits—with
a few exceptions, have had country-born mothers.   They transmit to their sons those traits of character—moral, intellectual,
and physical—which give stability to institutions and promote
order, security, and justice.  , When there have been remarkable
deviations from this law of descent, the mothers of city nota-
bilities, in whom talent has been the lever of eminent success,
had opportunities for alternating between town and the life-
bestowing country.

City-born women, affected by morbid desires and corporeal
deterioration, jealously reared within those centres of exclusive-
ness which know neither merit, accomplishments, nor respectability, not supported on all sides by golden props, cannot boast
of the superiority of their children.   An influx of pure blood
from the country, to replenish languid fountains in cities, is
the salvation of a family.

From whence came those ladies who are pillars in the temple
of Christian benevolence ?  From whence come the men of
mark in these United States ?  From the country.  None of
them were born in a brown-stone palace.   Such structures were
erected by enterprising capitalists who commenced their career
in market stalls, jobbing shops, before the mast, or possibly in
an oyster house.  A widow's son, or an orphan boy, who left
his village home in search of employment, are those who build
cities, control commerce, erect factories, sustain themselves in
places of honor, and are a credit to the age.


### DEVELOPING A SOUND FEMALE CONSTITUTION.

A decided way for receiving a permanent benefit from
mineral springs, is to visit them on foot, without reference to
distance.  Those who might receive some advantage from

reputed medicinal waters, are generally unable, from want of means, to remain, were they to reach them. Physicians assure us that poor women have fewer complaints requiring professional treatment, than those who are exempt from constant toil. They have occasional chronic ailments, and suffer from incidental exposures and accidents; but more women in comfortable circumstances are sick than might be expected, as an impression is entertained that domestic comforts are safeguards against indisposition. How much of it is to be charged to perverse habits, excesses at table, and a derangement of the system from having too much assistance in doing what they should have done for themselves, may be found in the writings of plain-spoken physicians.

Allow young girls free, open, out-door exercise in their pastimes and romping frolics, according to their disposition for such active gambols. Give them good, plain food, especially milk, fresh vegetables, and fruits of all kinds, in unstinted abundance. Do not limit their appetites. When they have had what their bodies require, they have had enough, but not before. In loose garments and opportunity for putting in play all their muscles, they exhaust their pent-up accumulation of animal spirit, which, if restrained by hackneyed old maxims,— that it is unladylike to be frank, spirited, and alive, they will degenerate into dawdling nonentities, who may have the forms of angels without their attributes.

Permit young girls, without reference to their age, to run through fields, climb over fences, swing under the trees, gather nuts in the forest, and pick berries in pastures, if they incline to do so. If they racket through the hall, overturn chairs or break broom-handles, in their innocent sports, they are laying a sure foundation for health, elegant figures, blooming cheeks, and brilliant intellects. That is the way nature proposes to fashion lovely woman. It is a course of discipline which brings out in rich perfection graces that no treasure could purchase.

Exercise which results from the pursuance of some kind of industry, above all, should be warmly encouraged, as being most conducive to the health of body and mind. A door that is always closed will have rusty hinges, and creak when opened. Indolence is an enemy to felicity. Keep busy, therefore.. A wise mother will find employment for her daughters. If they are idle, then they will be unhappy.

Were our limbs rarely set in motion, they would become thin and feeble. The skater's legs increase in size by the service imposed upon them. If the brain were not employed, poverty of thought would expose the neglect of that organ in exhibitions of ignorance. Decay follows neglect, and neglected opportunities cannot be retrieved. Every faculty must be exercised, if possible. Effort becomes a pleasure. Progress and prosperity have no intimate connection with pain or misery. Great thinkers, like fleet horses, must be kept in constant training. Great things are not achieved by main strength. Occupation is one of the first elements of happiness. As it is a woman's mission to smooth the rough ways of the world by the influence of her character, the power she wields is strong or weak, according to the culture she has received.

Commence seasonably, then, with young girls, by allowing them all possible freedom, not inconsistent with purity of heart. The best gymnastic school for them is all over the premises ; and when their bodies have taken the form nature contemplated in their organization—which is always beautiful —then teach them whatever may be requisite for sustaining themselves with propriety, dignity, and honor, in all the social relations to which they may be called.

# CHAPTER IX.

## NERVOUS SYSTEM OF WOMEN.

Different Nerves—Their Functions—Anatomically alike in both Sexes—Old Age—Children Nursed by Men—Arrest of Pulmonary Consumption by Lactation—Too Much Restraint—Exercise Essential.

FAMILIAR as medical gentlemen suppose themselves with the structure of the body, a woman's instincts, and the laws which govern her nervous system, are still veiled in mystery, which the acuteness of physiological research has not cleared from obscurities. Assumptions are not demonstrations. The curtain must be raised higher before we can explain phenomena which belong exclusively to female life.

Considered as an animal, man is not affected by revolutions of the sun, the moon, or planets, nor have conjunctions had any influence over his organization.

In women, on the contrary, there are periodical changes occurring with an orderly regularity which popular opinion supposes is wholly due to an influence of the moon, far off as it is. Before science had that ascendency it now has, there were such precise and inexplicable functions performed from puberty to about the fiftieth year, it was natural enough to suppose a power in the sky that made the tide rise and fall twice in twenty-four hours, also moved fluids, wherever they were, proportioned to the volume upon which the lunar influence was exerted.

Whether the nervous matter is fluid or solid, within the sheaths called nerves, is not satisfactorily settled. Those flexible white cords, from the size of a pipe-stem to filaments too attenuated to be seen without a magnifier, and which reach every fibre, being the telegraphic wires through which volitions are sent from the brain, and through which sensations from without are forwarded to the soul, have their origin within the head and upper part of the spinal marrow.

## THE BRAIN.

Nerve-cords are precisely alike in both sexes, have the same relative locations, and sustain the same office. A nerve in a female arm in no way differs in composition or in function from a similar one in a male arm; yet the brain of a woman differs from a man's, not in composition, nor in the proportions the white bears to the gray matter, so far as we can discover, but in its manifestations. There is a difference between the two, not at all easy of explanation. Side by side, detached from the skulls, it would be impossible to decide which was the male, or which the female brain.

Education is simply a development of the faculties; and when the process is conducted precisely alike for both sexes, there are manifestations totally different, which have their origin from impressions made exactly from the same sources. Therefore, there is a constitutional endowment: the why or the wherefore our philosophy fails to explain. Woman's instincts differ very widely from man's. She is naturally more reserved, more moral, and more sympathetic. Their thoughts, their dreams, and the activity of their imaginations, are certainly influenced by the same agencies which leave impressions on a man's mind; still she has neither the instincts nor the characteristic impulses of man in the concerns of ordinary life.

Whether the moon, the planet Neptune, the seven stars, or

the whole combined, govern the fluids in a woman's body, or unite their forces with those of the heart, it would not be wise to discuss. Certain it is, physiology has further room for explanation where there is both darkness and guessing, rather than light, in regard to the nervous system of women.

The structure and nervous expansion of slender twigs set the microscope at defiance. Their extreme minuteness cannot be followed, and, therefore, we must acknowledge our inability to pursue them.

### CURIOSITIES OF LIFE.

When the age of child-bearing is past, the milk ducts shrink and almost wholly disappear; but they may be revivified by simply manipulating the nipple occasionally a few days. The increase of blood is directed to the partially obliterated breasts, and the erectile tissue receives an increased influx of nervous exaltation. By allowing a nursing babe to draw upon the dried-up fountains, the functions of these organs, as in the vigor of youth, will be reëstablished. Should the powers of nursing be renewed and continued at regular intervals a few days, milk will be secreted abundantly. Children have actually been nursed in this manner by aged women, who were fully restored, in that particular, to the prominent conditions of maternity.

It will be conceded, therefore, that there are mysteries pervading the female system, when such phenomena are presented. Thus, through the reflex influence of extremely minute thoracic nerves, a lost function may be reëstablished. Glands which have been dormant for years—the sleep of old age—yield nourishment abounding in elements which are the appropriate food of an infant, out of which its solid body is built up in health, strength, and beautiful proportions.

Medical books furnish the case of a poor Italian who positively nursed his own infant seven months, on milk secreted in his own breast. For the purpose of quieting the starving babe, whose mother had just died, the afflicted father, unable to provide a nurse for the wailing infant, allowed it to nestle at his flat, hard bosom, which was instinctive on the part of the little famishing sufferer, where, finding a rudimental nipple, it was permitted to draw upon it without interruption. It quieted the screaming motherless babe; and the father, discovering that it was an easy method of procuring rest for himself also, offered the remedy as often as the dependent little one demanded it. To his astonishment, it was soon found that milk was there, and the child receiving actual nourishment. For seven months he officiated in the capacity of a wet-nurse and saved it.

Young heifers may become milch-cows, precisely in the same manner, by the efforts of a hungry calf. This has been resorted to for gaining time, rather than patiently wait a natural process. However, it ought not to be practised.

About forty years ago, a young baby in Massachusetts was accidentally the subject of neighborhood sensation, which would have been a valuable contribution to a medical journal, had it not been for a fear of damaging the reputation of both the living and the dead, because it would be difficult to make even medical men believe the possibility of what has since been fully established as a physiological phenomenon; viz., that lactation may be induced without being a mother:—

An accomplished young woman in that stage of wasting pulmonary consumption which indicated a speedy dissolution, such was the severity of her cough, and the copiousness of her expectorations, was residing in the family of a married sister, the mother of a babe she was trying with considerable difficulty to wean. Being advised to leave home a few weeks—it was thought the weaning might be more easily accomplished than

while she was continually in the society of her crying child—
the mother departed.

The first night after her departure, the moanings and un-
ceasing crying of the child had such a disturbing effect upon
the debilitated aunt, who could get no rest in sleep, that she
begged the nurse to bring it to her bed, suggesting she might
succeed better in quieting the poor thing, than the woman in
charge. By tender attentions, which in fact consisted in folding
it to her bosom, without particularly restraining its movements,
and falling into a slumber from exhausting efforts, the little
visitor found a pap. On awaking, and ascertaining that the
infant was industriously endeavoring to nurse, she removed it.
But its renewed screams induced her to take it back again, and
let it have its own way.

Thus, day after day, and nights particularly, the weaning
babe was hushed into sweet repose. But what was the astonishment of the emaciated invalid, to discover she was not only
relieved of some of her painful difficulties, the cough being less
severe, her appetite improving, and the child thriving on a full
breast of milk!

A mortifying discovery to the aunt was this secretion, and
that she was nursing a babe seemed miraculous. On consulting
her physician, a discreet, philosophical gentleman, he advised
that she should continue the course, it being evident she was
rapidly improving from a condition of prostration quite hopeless
before the baby was taken into favor, and suggesting the possibility of perfect restoration to health, if the pulmonary irritation
could be thus favorably subdued.

With encouraging prospects, and obviously improving
rapidly, an event occurred that blasted the expectations of the
medical attendant.

On the return of the mother, unprepared for such gratifying

prospects of a recently almost dying sister, and astounded at what was related of the child, it began to be whispered mischievously, by meddling village gossips, that the putative mother was not the mother, as had been supposed, but that the innocent babe, doing so well for itself, was actually the child of its reputed aunt!

When the cruel slander reached the ears of the sensitive patient, the shock agitated her almost beyond pacification by sympathizing family friends. However, she resolutely refused ever to receive the child again, much as she loved it. Arguments and appeals were alike unavailing, although it was plainly explained to her that a sudden suppression of the lactic secretion would seriously damage her case. Nothing could alter her determination. By the time the milk disappeared she was in the grave, a victim of a wounded spirit.

It is not certain that the young lady would have finally recovered, for, as has been already shown, where the structure of organs essential to life are destroyed, new ones cannot be regenerated. But violence of symptoms may be abated, and life very considerably prolonged, even when the lungs have been extensively ulcerated, and abscesses formed in the body of the lobes.

### Domain of Organic Sympathy.

There is an unfathomable sympathy existing between the pelvic viscera and the mammary glands; and because we cannot explain it satisfactorily, it is better not to dwell upon a subject of so much importance in the successful practice of medicine, which cannot, to any advantage, be discussed in a popular treatise on the laws of health.

With peculiar delicacy of mechanism, woman has also a corresponding nervous susceptibility. Her perceptions, her intuitions, and her moral tendencies are her own exclusively, and, though allied to those of the male sex, her nervous system is peculiar, and differs from that of man.

Men may be refined, conscientious, timid and retiring, but still fall infinitely below a woman in those attributes which give dignity, grace, and loveliness to her character.

Women faint more easily than men, and weep, too, artistically when occasions require it; but no familiarity with cruel practices, no outrages or wrongs are perpetrated so frequently in their presence, as to deaden their sensibilities to suffering, to appeals to the heart and their characteristic sympathies.

They recoil at the sight of blood, scream at the approach of a mouse, yet, in defence of their children, face the mightiest array of force with a heroism that death cannot invest with horrors of sufficient magnitude to divert them from maternal exhibitions of moral courage absolutely sublime.

Reason as we may, and rear arguments mast-head high, with an expectation of making the world believe phrenology is nothing more than ingenious sophistry, quite unsupported by facts, there is one circumstance obtruding just where it is most unwelcome to the opponents of that much-abused science, viz., that a large brain has more power than a small one.

## HUMAN HEADS.

Small heads, it is assumed, are never distinguished for generating great thoughts. Further, there is a consciousness in coming into the presence of persons with large heads and broad open countenances, that an engine or a battery, call it by what name we may, is inside those strong boxes, which are a power.

On the contrary, a pin-headed individual, whether man or

woman, whose cranium scarcely rivals a cocoanut in size, without breadth—there is nothing commanding about it which impresses us with a conviction of superiority in or about the individual.

That placidly received doctrine, that all enter upon the stage of human activity upon equal terms, and with equal aptitude for being qualified by education to act the hero or be a knave, according to circumstances, is more charming in a Fourth of July oration, than true. Cash and circumstances, especially the first, as society now stands, carries more votes than talents, and buys positions which modest merit could not acquire by the practice of all the moral virtues.

Some are born to command, as others are to be commanded. This is exemplified in every form of government, from the nursery to a throne.

Notwithstanding an array of reasons advanced for giving women political, and, indeed, all other privileges which men glory in possessing, reference is not unfrequently made to their mental capacity, genius, and other cerebral attributes. They are not exactly underrated or undervalued, but there is a mean attempt at both, when impudence passes for argument. That they are inferior to men, just because they have not their staunch bones or do not chew tobacco, is a slender cord for binding up absurdities.

That the skulls of women are smaller, on an average, than male crania, cannot be denied. But that fact does not necessarily imply an incapacity for high pursuits. If they are small, there is a compensation in the quality for what may be wanting in the quantity. There are neither ganglions nor nerves in one that are not also existing in the other. Science or education will ultimately demonstrate that a female brain has a capacity for the reception of any knowledge men may or should acquire.

Education is a miracle-worker, especially when it takes female pupils in charge. That common notion, that woman's sphere is one in which there is no need of knowing much of anything besides sewing on buttons, rocking cradles, or dusting furniture in the drawing-room, cannot have many advocates,—certainly, none of sense.

### RESPONSIBILITY SOMEWHERE.

Fathers, brothers, and husbands are guilty of a great wrong if they neglect to elevate woman to the extent of their pecuniary resources, in giving her all attainable advantages. Her mission on this fair globe is such that she must have intelligent training. All her faculties should be developed, and directed to meet the responsibilities of her position.

Women are under too much restraint. They have been guarded in selfish ignorance, till a common sentiment has crept into our civilization that they still ought to have fewer privileges and less freedom than men.

Extreme reserve, seclusion from avenues to a familiar knowledge of what is transpiring in the world in which they have a being, is making prisoners of those who contribute all that is really refined, elevating, and heavenly in our sojourn in life. Reserve may be carried too far, and freedom degenerate into vulgarity. That system, however, which inculcates self-respect, has intimately incorporated with it dignity of carriage, generosity of soul, frankness of manner, chastened by the highest sense of propriety.

Where there is too much scrutiny—too much fear of being too common—it is impossible to have a full gushing out of a woman's real nature. Contracted views, hypocritical reservations, and concealment of motives, are always referable to a famished mind. The brain must be exercised, or it will perish.

A woman's nerves are cords of a delicate instrument—a harp of a thousand strings—which will not keep in tune if rudely handled. Whatever may be a success in the primary education of boys, should also be adopted in the primary instruction of girls. Quite into their thirteenth year, they should stand upon the same neutral level. Whatever is proper for one, is equally of value to the other.

### Boys and Girls in Childhood.

Boys of a corresponding age, owing to their innate disposition to frolic, and who in their boisterous pastimes put in action every thread of a muscle in their agile bodies, invariably have larger, stronger limbs than girls. Besides the circumstance of having larger bones, too, males of all the higher order of mammalia possess an original endowment, in the general size of the whole body, above the female.

Were girls permitted to exercise as boys do, unrestrained by maxims and trammels which ignorance imposes at home and abroad, in the nursery and the school-room, they would become nearly as muscular, and much more expanded. Their chests would be broader, but an instinctive delicacy never forsakes them under the roughest usage, or the most vulgar, demoralizing associations.

### Over-Educating.

A mistake in female education that will have to be rectified, before women have their true position, is over-doing. They are educated too much. Their ductile minds are developed prematurely, to the positive injury of their bodies, before they have fairly begun to live. This, too, is all wrong, and one of the causes of nervous irritability and excitability peculiar to females in the Northern States. They are made learned, without being practical.

A discourse on astronomy, or criticisms on a musical composition, make an accomplished daughter. When she becomes a wife, she is at. the mercy of servants, and her husband quite undervalued, if he knows nothing beyond providing bountifully for the comfort, honor, and respectability of his family.

Thousands of ladies are too erudite to be of any use anywhere. They look with contempt upon those who have not had equal advantages for being made useless beings like themselves, and yet, when examined by the test of common sense, they have never contributed a new thought, or, with their accomplishments, enlarged the circle of human happiness.

It is not a crime to laugh at the rural habits of a plain farmer, but it is a disgrace to a fine woman to ridicule the simple manners of country ladies. If they could see themselves, occasionally, as sensible people see them, there would be a change for the better.

## PRETENSIONS TO ADVANCED KNOWLEDGE.

Some city ladies entertain exalted views of their own superiority over their country friends. When the well-meaning Mrs. Baker, the grocer's widow, retired with a competency, she purchased a pleasant domain of forty acres in the town where she was born, one hundred miles from the city. Her head was stocked with scientific agriculture, gleaned from the best treatises on farming made profitable. She had not been two weeks at the new home, which was undergoing astonishing repairs, before she discovered the extreme ignorance of her neigh-

bors in respect to rural pursuits. Not one of them had ever
read a page of modern agricultural literature,—fascinating books
for a cultivated mind. She resolved to revolutionize farming,
by showing the world generally what could be done by a city
woman with a will and money combined.

For the purpose of extreme accuracy, Mrs. Baker having
further resolved to report her successes in experimental farming, she had a leaf in her diary ruled off for profits. She explained, in the kindest manner, to her coarse neighbors—in her
opinion the most wooden-headed creatures she had ever known
in the shape of men—that each bean, before planting, should be
weighed in Professor Pollock's patent agricultural scales. In
that way an exact register of the increase could be determined.
Fertilizers, purchased in tin canisters, which could be carried in
a vest pocket, contained the virtues of a cart-load of nasty
manure. Instead of delving with a hoe to clear away weeds, a
pair of Sly's patent vegetable tweezers were worth a dozen hoes
—with that instrument the operator could extract weeds with
gloved hands.

All this was novel intelligence, really quite new to Mrs.
Baker's astonished auditors, who said nothing beyond expressing uproarious wonder that a great lady had known so much
about a subject they had generally supposed did not come under
the catalogue of book-knowledge. They noticed she expressed
herself in long terms, not in all the dictionaries.

It was revealed to Mrs. Baker that some of her most re-
spectful listeners, in appearance, actually laughed behind her
back. "Never mind," replied the philosophical reformer,
whose zeal had not degenerated into lunacy, "let those laugh
who win."

When harvest arrived, those ignorant farmers had excellent
crops without having consulted an encyclopædia, while Mrs.

Baker's manager gathered less than had been sown. She opened her eyes with amazement to the solemn realization of a singular fact, viz., that too much science is unprofitable, if one intends to live by farming.

Female education may be deplorably defective when women are taught too much of what is of no earthly value to them, at the expense of their health, and equally so, when they assume to know what they do not know. Their systems may be destroyed by over-taxing the brain, while the machinery of organic life, on which mental excellence depends, is considered either of secondary importance, or quite overlooked.

### UNSOUND WOMEN.

It is a national calamity that the women of this country are so generally unsound. Those distinguished for brilliant intellects are the most common invalids. To be under medical treatment is not only necessary, but very genteel.

A gentleman of ample possessions and of excellent social position, gave it as a reason why he did not marry, that he did not feel able to keep apothecaries and doctors continually under pay!

Women would not be so nervously excitable, slender, fragile, sharp-featured, and petulant—as too many of them are, for the happiness of their households—if they had not been wronged in the beginning, through a mismanaged education. They would not have been so universally predisposed to dyspepsia, neuralgia, paroxysms of depression—which throw a gloom through a pleasant home, and discourage indulgent husbands—had they been generously permitted to breathe outdoor air, subsist on plainer food, rise earlier, sit fewer hours at a piano, and read something superior to sensational magazines devoted to exaggeration, moonshine sentiments, love in a bower, and other kinds of nonsense calculated to mislead and over-excite their youthful imaginations. This is the misfortune of what society is pleased to call the better classes.

The best-informed young ladies—those whose educational advantages embrace whatever is supposed will fit them for the highest positions which refined society has at its disposal—are the poorest wives. Matrimonial disagreements and wretch-edness are not found in the middle classes, but just where the refinements of the lady of the establishment enable her to discern imperfections where she has fondly hoped to find a companion who would sigh perpetually, recite poetry, and buy cosmetics by the gallon.

Their petulancy, curt answers, despotic rule of servants, and dissatisfied expressions toward those who are devotedly endeav-oring to promote their happiness, cannot be cured by pills, soothing powders, strengthening plasters, annual jaunts to Saratoga, or the attendance of a high-priced doctor.

Liberty to exercise in childhood, without being constantly reminded that it is unladylike to run, vulgar to eat enough to satisfy a moderate appetite, and wicked to be natural,—but charming to cultivate hypocrisy, improving to be fastened in garments that restrain the growth of the chest, and glorious to be in misery for the sake of dying a real lady, is the lamentable cause of many of the common woes of elevated domestic life.

Men and women were designed for each other on the high-way of the world. They are destined to the same length of days; and, above all, it was not intended in the original con-stitution of humanity that they should be strangers to each other, unless formally introduced, after having carefully in-spected a pedigree.

## NERVOUS CENTRES.

Besides injuries originating in the vice of dress prejudicial to health, suggestions ought to have been made respecting the violence inflicted on nervous centres. Just under the line where the pressure is most severe in girding on the waists of dresses, are the solar and semilunar ganglions. They are way-stations into which nerves enter and others go out, which hold control over the stomach, liver, spleen, etc. They are the brains of the abdominal viscera. They surround a short, horizontal artery that shoots off from the trunk of the aorta, the great arterial tube from the heart. The cœlic artery—not over an inch in length—subdivides into three. One goes to the stomach, a second to the spleen, and the third to the liver. Any compression of the waist, therefore, besides disturbing those nervous centres, interferes also with a free circulation of blood to three important organs in the abdominal cavity.

Mandates, or volitions, are sent from the brain, but the way-stations—those ganglia—repeat the commands. Unimportant transactions, when everything is progressing in the usual way in the viscera, are not transmitted to the principal office,—the brain. When there is unusual disturbance, pain and inflammation, then word is sent forward, and the judgment determines how to act.

Simple irregularities of digestion may occur, but unless there is a grave condition of things, the brain has no immediate knowledge of it. It is not always necessary to communicate what may be transpiring in any one organ, unless its functions are seriously impaired: then a dispatch is sent upward to the brain.

There are many considerations connected with the subject of the nervous excitability of women, which have called forth expostulations, but to no purpose.

Were physicians to write plainer than ever, and address themselves to parents, instead of preparing elegant essays for journals, walled in by so many barriers of technical phrases, which nobody understands who is not an expert in half a dozen dead languages, no change of system would follow. Fashion is antagonistic.

## Social Relations of Children.

When boys and girls are brought up together—in large families, sitting at the same table, mingling in each other's society, sharing in amusements and intellectual pursuits—they invariably go forth with better principles, stronger convictions of what is duty, and live purer lives, than those who are taught that it is sinful to look each other in the face, unless in the presence of a watch-dog of a parent, or a dilapidated old duenna, whose eyes can be covered with a ten-dollar bill to oblige her young mistress.

## Important Caution.

It should be taught children, that the pit of the stomach, as it is called, is nearly over those ganglions, or nervous centres, and that they must be favored in swathing the chest. A blow there is almost instant death. Life explodes, as it were, by any rude approach. A kick of a horse, or the weight of an angry man's fist, at that spot, is almost invariably fatal.

There are numerous glands in the neighborhood of the ganglions, which compression disturbs, and any interruption in their appropriate offices affects the general health; especially those connected with the function of chylification, if pushed from their natural relations, or in any way interfered with. They may become scirrhous, enlarged, hypertrophied;

and a softening of the bones, too, is sometimes referable to a similar cause.

Mendicant children of both sexes, common in public streets, scarcely covered decently in tattered loose garments, which are the cast-offs from persons twice their size, are in robust health, with splendid forms, sound white teeth, thick hair, round limbs, and good brains for cultivation. The rich man's daughters are forced into being ladies before they know the meaning of the word, by a system of unnatural discipline that kills them by inches.

Excessive fear of mingling with persons with a small rent-roll, or with none at all, and harboring the opinion that men are monsters seeking whom they may devour, are productive of nervousness and feebleness, traceable to the present system of female education; and which has also immensely multiplied maiden ladies, to the detriment of their own happiness and the best interests of society.

# CHAPTER X.

## AMUSEMENTS OF WOMEN.

Young Animals in Sports—Blind Buffaloes—Reptiles—Brain Volume—Mechanical Ingenuity—Conversation with Children—Theoretical Schemes of Female Education—Dancing—Entertaining Distinguished Guests—Theatres—Always have Existed, and Probably always will—Labor—Children Overworked—Philanthropic Efforts—Playtime a Sanitary Measure—Why Sleep is Necessary.

No suggestions can be made, or plans proposed, for the innocent amusement of youth, that will not meet with opposition from some source.

The absolute severity of some parents, who believe they have the special approbation of heaven for making their children wretched by interdicting amusements, is very surprising, since, in their own youthful days, many of them were distinguished for reckless, rollicking lives. It can be explained on no other principle than by a common observation, that the greatest sinners become very exacting saints.

All young animals have their sports and festive gambols. It is a natural way of exercising muscles, while under the excitement of pleasurable emotions, to act as they were intended to contract and relax when matured. Thus, they run, turn short corners, and seize each other with a tender grip precisely as they will hold their prey when urged by the stimulus of hunger.

Such, certainly, are the characteristic manœuvrings of carnivorous quadrupeds. Cattle, sheep, horses, etc., till their teeth are fully grown, are particularly playful, when they become grave and cautious. The lion's whelps, young tigers, foxes, bears, and those of a similar type, are extremely playful while nursing; but as soon as their stomachs crave more substantial food, the ferocity of their nature is manifested. Puppies are very sportively inclined, nor do they express their canine energy till they have had a taste of flesh. They then begin to quarrel among themselves, on the slightest provocation, which terminates in terrific fights for the possession of a bone.

Grass-eating animals rarely give such vehement displays of irritability, even when goaded by pangs of extreme hunger. Colts, calves, fawns, kids, rabbits, etc., delight in the freest exercise of their limbs, if in sight of their mothers. The males only engage in combats.

When battles are suspended, and renewed at short intervals, it is solely for the purpose of recuperation. A contest, once terminated by the submission of one of the belligerents, suffices for the remainder of their lives. The victor ever after walks abroad in the consciousness of being without a rival. Stallions, dogs, and bulls, when once conquered, remain in subjection while the conqueror lives.

Blind buffaloes are actually leaders of immense droves, by virtue of their prowess in youth, which is respected by hundreds of brave bulls, stronger and younger, demeaning themselves peaceably in the herd while the acknowledged ruler is able to move.

Skeletons of bucks are often found in deep recesses of the forest, with their branching antlers so inextricably interlocked, that the combatants must have died, in that painful condition, of actual starvation.

In these illustrations of the youthful propensities and habits of animals, the law of might is allowed to predominate. There are no rights acknowledged among themselves. Puny, feeble, imperfectly-developed males cannot compete with the large, the strong, and the highest type of the race to which they belong. Consequently, nature secures, in perpetuity, all the best properties for a succeeding generation.

## WHERE THERE ARE NO AMUSEMENTS.

Neither birds, reptiles, nor insects appear to have amusements or periods of sportive relaxation. From birth, they are sedulously devoted to habits of industry, in providing for their own necessities and the wants of their offspring. There is an instinct always making reference to successors, but the sentiment of parental affection is not long-lived with them.

Pigeons and domesticated doves fly about in groups, foraging, but they never visit each other's cotes, nor engage in sports. The attachment of the parents, when once paired, might be advantageously imitated by reasoning beings, who find more relief in the laws of divorce, than comfort in dove-like matrimony.

Their attentions to their young are of short duration, and quite at variance with some other traits, which have been poetically lauded as worthy of consideration.

Fishes, crabs, lobsters, turtles, prawns, etc., seem never to have sports among themselves. Serpents, frogs, toads, and lizards are solitary as oysters, each intent on selfish pursuits. Whenever they do huddle together on the margins of pools or in cliffs of submerged rocks, they never indicate the slightest gratification, or hold any intercourse with each other, more than with other inhabitants swimming in the same element.

While young birds are being fledged in a nest, they lie quietly, without the slightest show of playfulness. Chickens, turkeys, goslings, ducks, prairie-hens, partridges, quails, peacocks,

and caged songsters, press together for mutual warmth or protection, without manifesting the remotest show of a disposition
for sport.   Of the attachment of mother birds, nothing can be
more demonstrative for a short period.   They brave all dangers for the protection of their little ones, and die in unequal
struggles for their safety.

Affecting scenes are described by arctic navigators, of the
attachment of polar bears for their nursing cubs, and the bloody
encounters they have been known to maintain against rifle-
balls, in unequal efforts to save the objects of their affection.
Usually, the male is rather an indifferent spectator.   Pairing
birds, and perhaps a few of the pairing quadrupeds, make some
show of interest in the young, in their most helpless infancy,
and join with the mother in defending the lair; but, as soon as
they are old enough to run and look out a little for themselves,
the father loses all interest in them.   Warm-blooded animals
are those which have pastimes, rude to be sure, but, nevertheless,
they actually enjoy social recreations.


### THE BRAIN.

As the volume of brain augments, a disposition for playfulness is more apparent.   Boys and girls scarcely do much else
from the cradle to adolescence, than play in some manner that
promotes their happiness.   The fabrication of toys of any con-
ceivable description for their amusement is a branch of manufacturing interest that has always been profitable, and gives
employment to large numbers of ingenious mechanics in every
country, civilized or not.   Very large commercial houses are
exclusively engaged in the importation of playthings for
children.

In the catacombs and mummy-pits of Egypt, and the cem-

eteries of extinct nations, toys have been found in abundance,
showing that the demands of childhood have always been recognized in every age and country where humanity has had a
being.

The disposition, tendencies, and irresistible demands of
their nature for objects proper for exercising their juvenile
brains, is a necessity, and has been, from the first formation of
human society. Savages tax their ingenuity in making rude
toys for their children.

This may be thought a small matter, but it is of importance.
Toys assist them in forming opinions, correcting their judgment, and in classifying muscular action. Distances, weight,
dimensions, form, color, etc., are insensibly acquired, to be
applied in other ways, and for far different purposes as they
advance in knowledge.

## Give them Facilities.

Mechanical skill and a genius for invention is very early
manifested in some boys. They should always be gratified with
the possession of implements for perfecting their designs. Too
generally they are denied facilities which would give them
great advantages. Tools are invariably coveted by such as have
a mechanical turn, but nothing is more common than to deride
their plans and ridicule their machinery. Let them have ham-
mers, saws, chisels, files, and by all means a turning-lathe, even
if they break some of them and lose the remainder. Children
have wants, real ones too, which, when not positively preposterous, should be indulged. It may lead to proud results. Every
one who has had experience with children, knows what a treasure a gimlet is to a boy. A jack-knife is something above
riches. With it he converts shingles into wind-mills, carves

horses out of turnips, builds edifices with blocks, and makes happiness for himself with it in a garret. With a box of tools he learns the use of instruments, while exercising both brain and muscles. With tools, boys can amuse themselves in the dullest weather. While mending their broken sleds, or constructing a miniature wagon, they are creating something, which is always a pleasure. The little miss, in dressing her doll, finds unspeakable enjoyment. It is teaching her how to use the needle, the thimble, and her scissors ; therefore, it is not a waste of time, but a regular course of instruction, in which practice makes perfect.

If girls and boys are benefited in no other way with tools appropriate for each in the sphere in which they have been designed to move, it is in being kept out of mischief while they are permitted to use them. They always love and honor parents who indulge them in the line of their social propensities. Constantly forbidding them to do this or that, because they themselves dislike it, makes disobedient children. If men and women are but children of a larger growth, they surely ought to sympathize with youth, and not exact of them sedateness, or the solemn expression of thoughtfulness that belongs to mature age.

## ASSOCIATE WITH CHILDREN.

Those parents who never allow themselves to mingle with their children, or express an interest in their little pursuits, have no foretaste of heaven. Conversation with them gives them encouragement in what, to their immature minds, seems of the highest importance. Ridicule is a hateful weapon in damping the ardor of ambitious children. Help them on with their inventions; assist them in their rude drawings; suggest improvements in their constructions; for a great architect, a

splendid artist, or a distinguished engineer may be hidden in
the rough combinations of blocks, old bricks, or snow-balls rolled
together at recess in front of a country school-house.

Young women, like' young men, must have amusements.
It is an inborn necessity of their nature, and hence the question,
What may they do or not do, after passing through rattles and
dolls?

A host of propositions emanate from all sorts of people in
regard to the question, which is thought more momentous in
reference to girls, than the question merits. There is a plain
way of settling the matter, in accordance with the acknowledged rules of Christian propriety and benevolence.

It is curious that more schemes for rearing young ladies to
be what society expects and demands them to be, emanate
from persons who never had daughters of their own, than from
those who have had many to perpetuate their memory. Let
them have the confidence and intimate society of their parents.
That is one of the first lessons for improving them.

Theoretical schemes on the culture of female youth almost
always have their origin with sour, opinionated old-bachelor
teachers, or, worse still, retired maiden ladies.

### INFLUENCE OF MUSIC.

Nothing more instantaneously quickens nervous excitability
than instrumental music. Some airs have such inspiration in
them that we can hardly control our feet, which is a very direct
mode of conducting off what the brain is taking in. If octo-
genarians unconsciously beat time with their gouty toes, what
electrical ecstasies get the ascendency of young ladies and
gentlemen, when the thrilling tones of a violin break in upon
their ravished ears!

Dancing is one of those natural, spontaneous outbursts of youth, which can hardly be suppressed by efforts of the will. It is not to be first learned before a disposition to dance is developed. Those who know nothing, artistically, of taking steps, or ever saw others dance, can scarcely restrain themselves from a sudden display of sprightly antics, when music rouses them to a state of exultation, which cannot be produced by any other means.

Why have we ears for music, or music at all, if it is wrong to listen to it? Why is Old Hundred any more acceptable to that Divine Intelligence, who is the author of harmony and the contriver of our acoustic nerves, than the College Hornpipe?

### DANCING.

Dancing is an admirable exercise for all the cordage of the body, and eminently conducive to health. It quickens the circulation, while promoting all the glandular secretions. Nothing else compares with that exercise. Nature intended it for a peculiar sanitary pleasure. Although young animals do not artistically dance, they caper and display their agility under the exhilirating excitement of exuberant health.

We dance to sounds that stimulate a more highly-organized brain than animals possess, till weariness succeeds, which is an evidence that no further excitation is required for that time.

Let hard-faced, dilapidated casuists reason as they may on the moral torpitude of dancing, it is perfectly in harmony with those hygienic laws, the observance of which, never carried to excess, tends to health and longevity. King David danced before the Ark of the Covenant, for which he was severely criticised by one of his pious wives, whose pharisaical hypocrisy was of a piece with the moral shock the sensitive objectors to a very innocent recreation pretend they feel when a ball is proposed.

Religious intolerance, from immemorial time, has been at open war with the votaries of the dance. It is, indeed, remarkable that the clergy of some denominations never fail, when opportunities present, of thundering anathemas against that odious so-called sin, as though it were a dreadful crime in the sight of heaven.

Under the shadow of those edifices where fearful denunciations are annunciated against that shocking vice, and where solemn pronunciamentos are regularly promulgated, dancing-schools flourish with undiminished success. Dancing has never been abandoned in any community where those great ecclesiastical guns have been levelled, nor ever temporarily suspended on account of the bigoted hostility of bilious sour-krouts, who are never happier than when they have made some lady wretched, in obedience to their interpretation of the Divine Will.

Government officials and municipalities greet distinguished guests with cordial attentions, which usually embrace festivities in which dancing is a prominent feature.

### Moral Sentiment.

Those self-constituted instructors in moral excellence, who presume to assert what is most pleasant and satisfactory to themselves as being also most satisfactory to the power above they represent, gain nothing for morality by their hostility to innocent amusements. Ecclesiastical cannonading avails nothing, since people will continue to dance while they have feet and music is heard on earth.

## THEATRES.

Theatres are universally denounced by the same self-constituted interpreters of divine precepts, as the focus of demoralization; but, notwithstanding the unrelaxing bombardment to which they have been subjected, they are multiplying with the extension of civilization. Delighted crowds throng them, and they will continue to do so, while society exists in its present form.

Were it true that scenic representations of the foibles or the graces of mankind on the stage were as bad as ranting reformers represent, a second deluge would have been required centuries ago, to wash away their pollutions.

Dancing, music, and theatres will be sustained while men have ears, music charms, and the stage represents the passions, hopes, fears, love, and hatred engendered in the human heart. No legislation could arrest either, or suppress them so effectually as that they would not reappear in some form essentially the same.

Appeals to the conscience have been as ridiculous as shooting at the moon with an expectation of forcing it from the orbit in which it moves. Persecution is ineffectual. When legal enactments are sustained by a force strong enough to stop public amusements, of which dancing and theatricals are most prominent and universal, because they are considered a nuisance or a sin, then moral reformers must interdict music also, in the same bill. After that, to be consistent, ears must be cut off, whenever it can be proved before an impartial jury of self-constituted saints, any man, woman, or child, wickedly, and with malice aforethought, listened to prohibited strains of melody, against the dignity and majesty of an offended law. There are in Europe, at the present moment, fourteen hundred and eighty-two theatres.   In France, three hundred and thirty-seven ; in Italy, two hundred and eight ; in Spain, one hundred and sixty-eight ; in Austria, one hundred and fifty-two ; in Prussia, seventy-six ; in Russia, thirty-four ; and in England, one hundred and fifty-six.   In the United States, where they are numerous and constantly on the increase, the Canadas, Mexico, South America and the West Indies, make a very formidable list for the New World.

### More Recreation Demanded.

The opposition which narrow-minded people manifest against dancing, is perfectly unaccountable.   In the New England States, there are not recreative amusements enough for the proper relaxation of body and mind.   Public sentiment was formed by Puritan ancestors, who were compelled to work incessantly for their preservation.   They had no opportunity for relaxation or social enjoyment.   Their ecclesiastical teachers, in whom they reposed implicit faith, and to whom they yielded servile obedience, were careful to instil into their crude congregations the heinousness of levity.   The wickedness of laughter, and blind devotion to the gloomy teachings of a church that fled from oppression to become an oppressor, was inculcated by saintly men who vigilantly superintended their flocks. Labor was necessary, but they were over-taxed with cares which gave a fixed gravity of countenance that has been transmitted to their posterity.   This accounts for the haggard, gloomy faces which predominate there to this day.   They are taught to do everything from a sense of duty, and never to allow any outgushing impulses of hilarity.   It is quite remarkable, that with the progress of society in art, science, literature, and humanity, there are still many remnants of the good old times referred to in their chronicles, who deem any deviation from their standard of faith, a near approach to an abyss of misery in the world to which offenders are hastening with railroad speed.

## Over-taxing Children.

Children are over-worked—far beyond their powers of endurance. It is discoverable in their imperfect physical development. With us, their brains are over-taxed. · Schools of every grade, from primary infantile to normal institutions, require too much. Under the impression they are having rare facilities for acquiring knowledge, the poor things break down under a pressure of too much instruction.

Force of circumstances compels parents to place their children too soon in factories, where they are wronged out of their share of vital air to which all are entitled. Philanthropists have appealed to the legislature, but in vain. There is law enough for their protection, without a corresponding earnestness to execute it. Though all are born free, and have equal rights in the pursuits of health, wealth, and happiness, only few of the many secure either. Poverty connot compete successfully with wealth.

There is another field for culture where the harvest might be large, but the laborers are few. In private families where children are loved and watched over with paternal solicitude, there is a culpable ignorance in obliging their little ones to do too much, under the mistaken idea of giving them superior advantages.

Precocious children disappoint the ardent expectations of their friends. When they arrive at an age at which they are fondly supposed to be ready to blaze with extraordinary mental brilliancy, their feeble light goes out. Slow and sure is a true saying. Gradually evolving an intellect, as a flower unfolds its beauty, is a safer process than bursting open suddenly, to wither under the first rays of a morning sun.

Children ought not to be taught much of anything more than moral duties, till they have reached at least six years. Their brains are in no condition for concentrating thoughts before. They should have perfect liberty to act out their exuberant playfulness with as little restraint as possible, consistent with proper discipline in the lessons of good manners, courtesy, truth, and order. Time is not lost in giving them such scope for exercising body and mind. Their activity and ever-varying amusements are but so many ways of tutoring their muscles, their organs of sense, and in preparing them for the places and responsibilities of the future.

Public schools are over-working pupils, goaded by fear of disgrace or punishment; over-excited by promised rewards, their immature nervous systems are forced at the expense of their vitality. When pale, delicate, frail little girls are flattered into a morbid ambition in a Sunday-school, to commit to memory long, dry chapters, to them without meaning, it is reprehensible. It is a violation of a physical law that has broken down and spoiled many a bright and promising child.

Allow children all the play-time they wish. They will stop at a seasonable period for disciplining their innate powers, voluntarily, to commence a higher series of employments which will be also enjoyments.

It is a lamentable mistake to keep young misses several successive hours at the piano. Dragooning them into accomplishments is a poor policy. Besides deranging the minute structure of the brain by long-continued practice at a single sitting, if attended with fatigue, the continued attitude presses painfully on certain bones. Curvatures of the spine, and a droop of a shoulder, are traceable to such circumstances.

Recollect the bones of young girls are not completely ossified till near their twentieth year. They are not hard and firm. A fixed attitude, therefore, so as that the weight of the body presses directly on the pelvic frame-work, may warp them out of the line in which they should have development. Nature has inspired all young animals with a restless spirit, on purpose to keep them moving. A love of change is simply giving each and every fibre and organ a chance to perfect its organization.

While children sleep, which is about all the rest their active limbs require, processes are then rapidly going on for the physical completion of their bodies. That is the reason why they require so much repose. Internal artisans then labor with intense energy while they are quiescent in slumber.

Growth is suspended when they are awake, but renewed the instant their eyelids are closed.

Unfledged birds in the nest sleep nearly all the time, after leaving the shell, till their feathers are sufficiently developed to sustain them on the wing. Their perfect quietude favors vital processes, so that in a very few weeks they are complete in all their proportions.

When the brain is large, the process of growth is slower. Allow young girls and boys as much sleep as they desire. It is not from indolence, or a sluggish nature, that they are so uniformly disposed to drowse to a late hour in the morning. If they retired earlier, they would rise earlier. But Nature demands both time and opportunity for completing their bodies according to a prescribed pattern. If we interfere with that law, and interrupt processes instituted for that purpose, they will have unfinished bodies, weak brains, and poor health.

# CHAPTER XI.

## Their Mode of Living.

With digestive organs requiring the same kinds of food that instinct and custom sanction for man, there is a special reference made in favor of some women, on account of a supposed delicacy of constitution. They imagine they could not subsist on ordinary diet. What they have must be very concentrated, so as to occupy but little room in the stomach.

Unfortunately, it is ungenteel to have much of an appetite, especially for young misses, destined to circulate in fashionable orbits, whose ignorant mothers commence early with giving them practical lessons in personal elegance. To dine heartily would carry with it an extreme air of vulgarity: hence, the less a young lady takes at table, the higher her preparation for refinements that are appreciated among those who think more of a fine form than of intellectual accomplishments.

Light soups, rich cakes, choice fruits, and tea always, is held to be the dietary range of an exquisite woman. Articles that would meet the requirements of her system are quite inadmissible, at least in the presence of satirical judges of propriety.

Food most approved, and that which carries with it the endorsement of manœuvring mothers, anxiously looking forward to the establishment of their children in commanding social positions, even if the intended husband is a baboon, is a slice of dry toast, weak black tea, and an occasional tea-spoonful of sweetmeats.

## HORROR OF FAT.

No calamity is more dreaded than fat in an aspiring young lady. Consequently, on the presumption that partial starva-tion is the legitimate way of keeping it at bay—that horrid destroyer of female symmetry and female ambition, of which very many are in painful apprehension—no efforts are left untried to preserve a slender form.

There are two methods extensively in repute for keeping off the enemy, which marketable belles manage with dex-terity. One is vinegar, drunk often; and the other, pickled cucumbers.

Those in comfortable circumstances, unsophisticated in the ways of acquiring extra attractions through the resources of art; those under no restraints from a dread of fatness; who satisfy a normal demand of the stomach, and breathe and exercise in an uncontaminated atmosphere—happily are re-moved from the temptation, the trials, discipline, and excitement of artificial life. But they are commiserated on account of their robustness.

Gaudily-dressed butterfly-misses, who are on exhibition in the street, at eclectic churches, if the weather is favorable for the display of feathers, diamonds, and streaming ribbons, are most frequently addicted to the vice of vinegar-drinking. A dread of fat is a misfortune, when it degenerates into an insane determination to be the shadow, rather than the sub-stance, of a live woman.

The consumption of pickles gives employment to many hands, and hundreds of broad acres are annually planted with cucumbers, to meet the mercantile demand—the consumers being principally ladies.

Gardeners and dentists are benefited by a trade that enriches both, while the effect is directly opposite on the health of that order of patients.

This is the country of poor teeth. A full, perfectly sound set is an anomaly. There are many with beautiful teeth; but there are ninety in every hundred young ladies whose teeth are in a hopeless condition of premature decay. Brush and cleanse them as they may, the progress of caries cannot be arrested.

No doubt, the quality of their food may have some influence in injuring them; especially, if taken either too hot or too cold. But large numbers inherit a predisposition to an early crumbling away of the enamel, which exposes the bony part to the direct action of agents that blacken and destroy the entire body of a tooth thus denuded of its protecting covering.

This diathesis is propagated and shows itself from one generation to another. Sound teeth, strong enough to resist influences that act unfavorably upon others, are also an inheritance.

Where an early predisposition to decay is recognized, there is the more need of supplying in food those materials which are appropriated for those organs in their growth, as well as preservation. With that tendency, acids hasten their destruction.

### COMMERCIAL PICKLES.

Pickles are but vehicles for carrying acids, and, hence, those who consume them excessively, especially those with an hereditary tendency to premature decay, quicken the process of decomposition.

Pure apple vinegar, or that manufactured from wine, is slower in its action than commercial vinegar, which is made of sulphuric acid. When diluted, it seizes upon the lime of the teeth with such activity, that the enamel gives way to its intense chemical agency.

Cider vinegar is too expensive for manufacturing pickles on a large scale. Sulphuric acid, therefore, is the basis of that of which common market pickles are made. It is not uncommon to find a cask of pickled cucumbers converted into a thick, pulpy mass of green gelatinous material, without any remaining resemblance to the vegetable from which it was formed. If too strong, this result is to be expected, kept barrelled eight or ten months, without being opened to the air.

Pickles, therefore, made from that acid, cannot be brought in contact with the teeth without doing an injury. Thus, in the expectation of preventing grossness, which, no doubt, is partially accomplished by acids, aided by a spare diet, caries and toothache may be anticipated.

## SOUND TEETH.

Travellers comment on our national tendency to defective teeth. Bad teeth, however, in the country, are not so common as in cities. There the food is not seasoned, usually, so highly, and is, therefore, freer from elements that undermine them.

In new countries, especially in wheat-growing districts, where lime is largely combined with the soil, men and women are tall, and the females particularly noticeable for their symmetrical proportions and admirable teeth.

Tennessee and Kentucky are celebrated for their splendidly-

developed specimens of humanity. Their out-door exercises
and plain fresh food provide nature with materials for completing her labors according to established laws.

When the soil is poor, thin, and barren of bone-making
constituents, the people are short, broad-chested, with lower
limbs disproportioned in length to the superior parts of the
body. There are tall and short persons everywhere, in every
community; but the average height is below that of the
inhabitants of places where the composition of the soil favors
their development to the utmost limits of the law of growth, as
in the new Western States.

Lime is scarcely appreciable by chemical tests where some
cereals are raised successfully, and where families are remarkable for their strong, fine teeth. Yet there are those among
them who have decayed ones; but the majority are favored
with sound, well-formed teeth.

Vermont, New Hampshire, Massachusetts, and the adjoining Northern States present an illustration of this fact in
regard to bad teeth bearing a certain relation to the agricultural resources of the soil. There wheat cannot be raised as
at the West, and there dentists are required. They are almost
as numerous as physicians. Dental operators appear very much
disproportioned to the population of the cities and towns in
which they settle. But that is accounted for in the facilities by
railroads for their customers, who reside in the interior.

Dentists are multiplying in the Western States, where once
the profession was hardly known. Their patrons are representatives of the Eastern States, in large proportions,—emigrants
from the worn-out, exhausted soil of the Atlantic States, who
carry with them the hereditary tendency to an early decay of
their teeth.

Estimated by the good they do in a sanitary relation, dentists are eminently entitled to all the honors and pecuniary independence they secure. When teeth, provided by nature, fail prematurely, art furnishes substitutes equally useful for mastication and speech.

The ingenuity of American dentists is not surpassed anywhere, in meeting the difficulties that present in thousands of irregularities in the jaws of the toothless.

Cereals are most abundant in phosphate of lime. Indian corn is not to be despised or underrated as food, because it is deficient in certain elements in larger measure in wheat. Wherever that grain is used extensively for food, good teeth are in the majority.

With the loss of teeth, not only the voice is considerably modified, but less distinctly articulated; certain sounds, essential to the perfect enunciation of language, cannot be given without them. Deprived of teeth, the expression is deranged. By a loss of the incisors, the mouth is out of shape, only to be restored by the substitution of artificial ones.

When teeth have been long removed, an absorption of the gums invariably takes place, which brings the lips together, shortening the face, and very much altering it—giving an appearance of age. When the original level of the gums is restored by art, sunken cheeks are again distended, and the muscles of expression immediately bring back the original characteristic outlines.

Because millions of teeth are blackened and eaten away by sulphuric vinegar, with a view to perfecting the form of the lady, by removing or preventing a superfluity of fat, pains and penalties, disastrous to the teeth, have been dwelt upon with a hope of wakening those, who are blessed with sound organs, to the nature of the disaster, the evils of which they may avoid by abstaining from factitious vinegar, and, if they can be persuaded, from every kind of pickle.

## UNBOLTED FLOUR.

Good flour, that most esteemed on account of its whiteness, is the poorest for food in those qualities which furnish tooth-matter. In bolting out the bran, there goes with it the materials indispensable for the formation of bone, and particularly teeth. Those, therefore, who subsist on coarse brown bread, made from unbolted flour, take into their stomachs precisely those elements that another class of good livers exclude, and they consequently have strong teeth and strong bones; while those, whose bread is of the finest and whitest quality, with their aching teeth to be filled or finally extracted, are the best patrons of dentists.

When Graham bread was introduced, a dietetic reform was needed. The bread in general use among good livers was too much concentrated. The flour was deprived of parts that should accompany it, in order to give distension to the stomach and bowels. The Graham flour retains the bran—the very thing of all others in the composition of wheat, which contains the phosphate of lime. When stablers feed their horses on that article, they give them something far better than flour. It is providing them with materials of keeping not only their teeth, but their bones, in good condition.

Ladies ordinarily subsist on food too concentrated. That is, it is too fine, and, therefore, does not distend the stomach enough to keep its walls from coming in contact,—a cause of many forms of indisposition, to which the poor, living on coarse, bulky food, are rarely predisposed.

## DIET.

There is a medium course to be pursued in diet, which entails no disasters, but favors health and exemption from incidental indispositions, that oftener have an origin in strong coffee, strong tea, and fine flour, than from any other cause.

Too cold or too hot are extremes in taking food. By cooking, all food is not only softened, and therefore made easier for digestion, but it destroys, by frying, baking, stewing, etc., parasites which abound in meats, fruits, and garden vegetables. Their eggs, too small to be seen without a microscope, are spread over and through almost every edible from the market by millions. Savages who take their food raw, or in a very crude state, are subject to a variety of intestinal difficulties. But their white, even, sound teeth show that they never have been subjected to the destructive action of hot drinks, concentrated acids, or beverages, which attack the enamel.

Perhaps the characteristic ferocity of savages is due to an almost exclusive meat-diet. Fishing and the chase, for supplies, is their principal employment. Fruits and vegetables are uncertain resources. Those they have are usually of spontaneous growth, with the exception of Indian corn, which is never cultivated in sufficient amount to supersede the necessity of ranging the forests for wild animals.

## THE METHOD OF LIVING.

Having explained the dangers to which young ladies are exposed, who deal too freely with vinegar, we now proceed to the consideration of the true way of living, for securing sound health and beauty of form.

Fish is both wholesome and nutritious. From very respectable authority it has been taught that the brain is especially benefited by it. Whether iodine, phosphate of lime, or pure phosphorus is taken from it by the absorbents, and carried to that particular organ, requires more decided evidence than has yet been adduced.

In the history of fisheries, fishermen have been distinguished for their bold, hardy, adventursome spirit, good nature, and indomitable force of character. They brave storms, breast dangers of the sea, and in ships of war, their spirit, gallantry, and reliability are acknowledged.

Mountaineers are another representative class. They are lovers of liberty, fearless, and the best of soldiers. Fresh air, plain food, and few wants, easily supplied, are excellent founda-tions for a vigorous constitution and mental activity.

Food has very much to do in the formation of character. With a strong, well-developed body, there is usually a corre-sponding spirit. A purely vegetable diet is not conducive either to a sound body or an active mind. Starch-yielding roots, as potatoes, arrowroot, etc., will support life, but they fur-nish neither corporeal nor mental power. Combined with ani-mal aliment, corn, wheat, barley, beans, fresh or dry, etc., furnish just those elements required in temperate zones for developing the best intellectual and physical capabilities of man.

There is neither strength of body, nor vigor of mind, when an individual is kept upon one article of food long enough to be loathed. The stomach must have variety, out of which are taken those substances required for keeping each and every organ in working condition.

Each particle elaborated by the vital chemistry of the digestive apparatus, is carried to the place required, as attend-ants on bricklayers transport mortar to the spot where brick is to be laid. When the new particle arrives, absorbents carry

away an old one which had been in relation with others a
sufficient time for imparting its specific vitality. As soon as
that has been extracted and appropriated, another should
arrive to take its place.

## LAW OF ASSIMILATION.

Thus the body is constantly undergoing a change. We are
reconstructed many times in a single year. Even the solid
bones are gradually removed, particle after particle, so gradually
and cautiously, that the fabric is neither weakened nor left exposed to dangers on that account.

Many times in an ordinary life of seventy years, the skeleton of every one reaching that age has been repeatedly renewed. This perpetual removal and introduction of new
materials explains the rationale of eating and drinking. It is
simply furnishing a crude mass, from which are selected such
parts as can be introduced into a living system vitalized and
assimilated.

A custom prevails of serving rare or uncooked meats, under
an impression that they are more easily convertible into nourishment. If cooked too much, the quality is imagined injured.
Thus, underdone expresses a condition that favors digestion,
while overdone means that it is not readily dissolved in the
stomach, and, therefore, is not as nutritious.

Neither extreme expedites, or essentially retards, digestion,
since the solvent properties of the gastric juice act with equal
potency on either. By habit, if a person has been accustomed
to hard-cooked meats, the stomach is prepared to receive that
kind of preparation, and that which is rare would not be acted
upon so readily ; and *vice versa*.

Soft-boiled eggs are usually served, because a notion is

extensively entertained that the stomach sooner reduces them to chyme. But a hard-boiled egg dissolves just as quickly, and yet, the egg-eaters are astonished at the suggestion that it is of no kind of importance whether eggs are hard or soft. Either way, they are quickly disposed of in the interior of that marvellous organ,—a human stomach.

Civilization, among other advantages over barbarism, requires that cooking should modify articles of diet. Cooking, too, destroys parasites which infest almost every thing in the catalogue of food. When introduced alive into the alimentary canal, the consequences are graver than when their ova are swallowed, which may not remain long enough for incubation.

In raw food, especially meats so rare as to be hardly warmed through, eggs of the tapeworm and the trichinus are actually introduced into the system. Rare meats, therefore, are objectionable on that account. Well-cooked food is safest.

## BUTTER AND SUGAR.

Butter contains materials for the reparation of teeth. Children are notoriously importunate for. it, urged on by instinct, too frequently interdicted by model mothers on the unfounded presumption it is too hearty for them. That it spoils their teeth and their complexion, are reasons given for denying it to them.

They would have better teeth for having as much butter as they desire. Egyptian taskmasters were told that it was impossible to make brick without straw, and it is equally difficult to have good teeth without phospate of lime, which belongs to the composition of butter.

Sugar, too, is usually withheld from children, who invariably crave it in far larger amount than it is given them. No demand of the system is more urgent than a desire for sugar. It cannot be overcome. Locking pantries, threatened punishment for invading sugar-bowls, never overcame the relish for it in small children.

The body requires sugar, and it must be had from some source. It is provided liberally in a mother's milk for her nursing babe. While thus fed, the infant is plump, round, and certainly lovely. When weaned, its dimpled cheeks fall away, the fat limbs lose their form, diminish in size, and the whole figure becomes more muscular.

Put upon a new diet, the quantity of sugar is much less than they had been receiving from a maternal source. So imperative is the appetite for sugar, that if not supplied in their food in the quantity required for the purposes of nature, a sugar mill is set in motion in the abdomen of land animals, and especially so in ourselves, to make up for the deficiency. This is one of the curiosities of organic life.

Every animal requires sugar. Some in larger quantities than others, but not one of them can do without some. Grass, hay, grain, rice, potatoes, beets, carrots, etc., contain it. By chewing and mixing with secretions in the mouth and throat, food is prepared for digestion; and when the essential properties finally mix with blood, there is extracted from it sugar.

In the liver, dark venous blood is redistributed. While passing through the vessels, there is extracted from it bile. This was long supposed to be the specific office of the liver. But it is made certain that the liver is a sugar-mill also. It supplies sugar rapidly, and the quantity made in a given time is perfectly amazing.

Whenever the supply is not equal to the requirements of their system, the deficiency is made up by a more active elaboration of it in the liver. It is necessary, therefore, and must come from some source, or a disturbance in the balance-wheel of life will soon be perceived; but, unfortunately, the true cause of waning health under such circumstances is not often understood.

Food should not be too compact. It is really an important point that it should have bulk enough to distend the stomach. The stimulus of distension is a condition required in the economy of life, because it facilitates digestion. That solvent fluid, the gastric juice, oozing, as it were, copiously from the lining membrane of the stomach, cannot act so advantageously on its contents in a fine, compact mass, as when loose and more readily permeable. Maceration, that is, simply being in contact with that secretion, is not perfect digestion. When there is bulk sufficient, at least, to press the membranous walls of the stomach asunder, it quickens the muscular fibres to contact, which rolls the ingesta from one part of the sac to the other, and thus brings new surfaces to the more direct action of the solvent.

### HEALTH OF LABORERS.

Laborers, sustained on coarse nourishment, have far better physical development, more strength, richer blood, and a far higher condition of health than their opulent employers, whose tables are laden with delicacies their servants and dependents may never have had the gratification of tasting.

Neither horses nor cattle can be sustained on concentrated food without seriously injuring them. Carnivorous animals have more compact aliment, but in them distension of the stomach is requisite for successful digestion. Feeding oats, barley, or any other grain to horses, exclusively, would soon be followed by gastritis, that would terminate fatally. The walls of their stomach must be alternately distended and contracted, to keep it in working condition. Meal alone, without hay, husks, or some equivalent, would not sustain a cow or an ox.

A dog, imprisoned in the cabin of a cast-away vessel, which floated about at random, after being abandoned by the men, was found alive on the twenty-third day. Although the poor creature had not a particle of nourishment in all that time, his life was preserved by the thick covers of a Bible, which he gnawed ravenously. But they afforded no nourishment. He lived on his own fat and marrow, which kept the lamp of life flickering, while the stimulus of distension which the Bible covers provided saved the prisoner.

## A REFERENCE TO CONTINGENCIES.

In the anatomical construction of animals, it appears as though a special reference was made to a possible contingency, in regard to a temporary supply of nourishment, by filling hollow bones with marrow, and cavities among muscles with fat. This is more marked in some than in others, which really seem to have had in view the possibility of the danger of starvation in the circuit they were predestined to act. Thus, a camel's life is considerably prolonged in their dreary voyages through deserts, where neither food nor water can be procured, by the absorption of fat from the hump on the back. Their ability, too, for carrying a supply of water that serves from ten to fifteen days, is an illustration of the fact, that an animal may temporarily feed upon itself, till relief is found.

Birds die of starvation sooner than quadrupeds, because their bones being hollow for the purpose of being filled with air instead of marrow, are not storehouses against a time of need, as in the other case. The buoyancy of feathered bipeds is due to the long bones being filled with air. They communicate with the tip of each and every quill, so that the barrels of all the feathers are filled with it also. If marrow was there instead of air, they could not fly. They could not have the same aerial freedom and levity.

Warmth of the body rarifies the air thus inclosed, and with motion the temperature is raised, which still further rarifies it, so that the longer they are on the wing, the easier they move. A wild goose is said to fly more easily the second day, on one of their semi-annual migrations from south to north and back, than when the jaunt is commenced.

There is a designing Power recognized in all these varied provisions for the preservation, not only of individual life, but also for the perpetuity of races.

## RECRUITING CITIES.

Cities are largely recruited from the country. New-comers arrive in the freshness and earnestness of health. They leave homes where they breathed a pure atmosphere, and subsisted on plain wholesome food. With what is conceived to be a bettering of their circumstances, on changing locality, they indulge in seasoned dishes, anomalous soups and delicacies quite unknown to the family from whence they came. A morbid taste is soon engendered, which craves repetition, till the rosy-cheeked clerk, or the blooming young lady, transported from a residence in a distant village to become the presiding goddess of a palace, have uneasy sensations. Their conversation is principally devoted to the discussion of what is good or bad for digestion, and they soon begin to discourse upon what may or may not be eaten with impunity. Next, medical advice.

There follows a physical deterioration of women, on their

transference to cities from rural homes in the country. When they pass half the night at an opera, dine near their original bed-time, admire champagne as a beverage, taking no open-air exercise, except in a carriage, formerly enjoyed on foot over green fields, chatting with pleasant, unsophisticated neighbors, as lovely as themselves, they fail. At last, with all their prosperity in social position, they are changed into pale, sickly, feeble fashionables, whose fingers, once round, full, and flexible, are reduced to the appearance of birds' claws. Sparkling diamond rings are not an equivalent for what they have lost. Artificial teeth, and perhaps a wig, made of the hair of some poor wretch who sold it to keep from starvation, shows what influence city life may have in the transformation of a beautiful woman to a pining, complaining, sickly lady.

Should they become mothers, their children are direct inheritors of their infirmities, the penalty of irregularities not catalogued as dissipations, but which are conditions invariably resulting from violations of the laws of health.

Without dwelling on the importance of abstaining from highly-seasoned, concentrated aliment for young ladies, it is obvious that they would have finer forms, health, and, consequently, brighter mental development, by subsisting on plain food. It is surprising that parents cannot be persuaded to adopt a system that promises, with moral certainty, to secure for their daughters sound health, the foundation for happiness.

Reformers are pointedly severe against some of the courses which we maintain are to be encouraged in the rearing of young girls. They are opposed to many exercises which are not associated with some kind of productive industry. In their shortsightedness, they discover no utility in a simple promenade for exercise, unless a miss is armed with knitting-needles, or is reading some solid work—like their own stupid productions.

Street yarn-spinning is not sinful.  It is profitable to walk the streets and see new objects.  While all the muscles are in play, shop-window impressions call into action all parts of the brain, which is as necessary for its good condition, as wholesome food is for the stomach.  If the eye is continually greeted with the same objects, or the ear has only a repetition of one class of sounds, neither of them will have the power they would have by varying both sights and sounds.  If the brain is always acting in one direction, millions of its fibres are lying idle.

The whole of us, as often inculcated in this work, must be used, or we cannot be developed into what we may have been.  Notwithstanding the severity with which street-walking for exercise is treated by those who cannot appreciate the value of it in training body and mind, streets have not been depopulated by the cogency of their arguments. Streets of cities are inviting pathways, and those windmill warriors who have discovered that woman's appropriate sphere is in the house exclusively, wlll never succeed in debarring them from the benefits to be derived from spinning street-yarn.

Exercise is so essential, it should always be encouraged, especially on foot.  When neglected, medicines and professional attentions are pretty sure to be in request.  Sedentary employments, or no employments, are equally pernicious, and are certain to be followed by derangements which would not have occurred had there been a sufficient amount of labor for the locomotive machinery.  Simply passing through the air, reclining at ease in a carriage, does not meet all the requirements of a living being.

## INCREASE OF RENAL DIFFICULTIES.

Renal complaints increase in proportion to the neglect of moving about on foot. The kidneys have a special office assigned them, of selecting out of the blood whatever is unsafe to be circulated throughout the system.

There is but one direct and prompt way by which noxious elements taken with our food can be conveyed away and thrown out, the retention of which would be injurious. It is, first, to dissolve them. Mixed with the chyle, they are thus introduced into the circulation, and sent to the kidneys. Whatever ought not to go further is intercepted by them, and separated from arterial blood, to be conveyed to the bladder, from whence it is voided. Therefore, the function of the kidneys is to be always in action, and never at rest. As the blood never ceases flowing in the vessels while there is life, the kidneys are alaways laboring without cessation.

By free foot-exercise the kidneys are very much assisted in their labors. Indeed, all secretions and vital processes are facilitated by it. Excessive indulgence in all malt or spirituous liquors, which are but too apt to stimulate the organs unduly, or, indeed, in any of those beverages palmed off on the unreflecting public as genuine, although really poisonous imitations, is very often the cause of a diseased condition of the renal glands.

The kidneys are vigilant sentinels that never slumber on their post. They carefully separate and send away that which would positively lead to derangement of other functions, if allowed to remain unseparated. Even when collected in the bladder, the necessity of relieving that receptacle soon becomes urgent; showing that what it holds, even thus secured, cannot be retained there more than a few hours, without producing immense disturbance.

Over-wrought brains, like over-worked kidneys, might have been avoided. Abstaining from drinks that excite the kidneys to excess, is an indispensable condition to the health of those organs.

A repetition of the lesson we are desirous of inculcating is pardonable, on account of its importance to youth. Simplicity in diet,—that is, plain wholesome food, thoroughly cooked,—is best for young girls, because it will secure for them a sound frame, and a clear intellect.

If they would adhere to those early habits which are usually customary in the country, after a removal to spheres of excitement, characteristic of what is thought, unfortunately, to be elevated social relations, they would be incalculable gainers. If they expect to escape neuralgic pains, a sallow compexion, loss of hair, decay of teeth, a wrinkled brow, waning vision, yellow moth-blotches where formerly there were tints of beauty, they must avoid the causes that produce those dreadful misfortunes. Whatever vitiates or impoverishes the blood, or over-excites the brain, diminishes the capacity for rational enjoyment; and a weak body, a debilitated mind, and premature death are the penalties annexed to the violation of the ordinary laws of health.

# CHAPTER XII.

## How They Should Sleep.

LITTLE or no thought is bestowed upon sleep, although a condition necessary for the health of every animated being. Man sleeps; beasts, birds, reptiles, fishes, insects, and even plants sleep. It would be quite impossible to live without it. While unconscious and in perfect repose, a recuperation is going on, rendered necessary from fatigue and a waste of vital force expended while awake. It is only in sleep there is a perfect recovery of something of that essential part of life which has been lost.

How or why we sleep excites no particular attention; but the place where repose is sought, and its surroundings, is an important subject for consideration.

Nearly one-third of the allotted span of existence is passed unconsciously, with closed eyes. The body should be in a horizontal position to have the full benefit accruing from it.

The lower animals sleep rather more than one-third of their lives. Reptiles, according to the climate, even more. Hibernates, in a sort of apoplectic repose, sleep heavily in northern latitudes, three or four months in succession. Alligators have a long slumber in the mud through a season most favorable for maturing their eggs, to be extruded on the return of a genial vernal sun.

Insects have their period of sleep, as profound, while it continues, in a house-fly, as an after-dinner nap of an alderman who engorges himself at the expense of his fellow-citizens.

## SOMNAMBULISM.

Somnambulic unconsciousness is an irregular working of the brain, which calls muscles into orderly action without the controlling will-power, necessary for conscious relations to time and place, for the security of the individual. Volition is partially suspended, and yet acts are performed while in that anomalous state, which so nearly approximate true volitions, that it perplexes philosophers in their attempts at a rationale of what transpires during its continuance.

Occasionally somnambulists perform extraordinary feats of daring without knowing it, or rarely having even a dreamy, confused recollection of what they may have done during a night ramble, in safety, where they would have feared to tread in waking hours. This happens, as frequently as otherwise, in the darkest part of a moonless night.

There are cases on authentic record in which a lady carried a lighted candle, and cautiously walked over a rapid stream, where she would not have dared to venture in full possession of her senses.

Some faculty of the brain, yet to be discovered, is in action during such exhibitions. Vision guides the footsteps of the somnambulist through dangerous passes, and the motor nerves obey the commands of the encephalon. When locomotive muscles receive a message, another set of nerves express back to the central station, within the skull, the reception of the order, followed by the required movements. All this transpires without consciousness, as though an independent mind were directing the machinery while the other was slumbering.

The immortal, indestructible entity—the soul—also reposes. This is inferred from the simple fact that unconsciousness, really and unquestionably, is sleep. We are obliged to express that condition by the word *sleep*, since no better term can be found that carries with it a more comprehensive meaning.

## ORGANIC LIFE.

While the organic mechanism by which life is sustained remains unimpaired, the current of vitality flows on uninterruptedly, indepent of volition. To a limited extent, it is quite beyond its control. Thus, the heart beats from the first moment of fœtal existence, months before birth, till the last expiring breath, not unfrequently one hundred years; and through that long period no effort of the will can arrest its pulsations.

We may hold the lungs from inflating by expelling the air a few moments; and, by practice, pearl-divers suspend respiration an incredibly long time, but vital necessity obliges them to come to the surface in about five minutes.

In sleep, the mind has no directing influence over the inflation or expulsion of air from the lungs. The circulation of blood, the contractions of the heart, and the return of venous blood from the extremities for revitalization, cannot be checked or accelerated by will-force. A sudden surprise, painful intelligence, or pleasurable communications, however, singularly quicken or retard arterial action.

Neither the heart, stomach, kidneys, nor any of the glandular bodies interspersed through the abdominal cavity, are supposed to have any rest or suspension from labor. They work continually without relaxation. Muscles, on the contrary, must have rest. The brain must have relaxation in sleep; and the soul, too, if confidence is to be placed in the deductions of science, demands undisturbed periods of repose.

In dreams, the mind does not have perfect repose. It is not refreshed under a state of emotional disturbance, and, hence, we complain of not having had a refreshing slumber. If the mind is not as completely quiescent and oblivious in sleep, as the voluntary muscles which it controls, then it is but imperfectly recruited. Long-continued seasons of imperfect sleep lead to grave consequences, such as impaired vitality, nervous debility, and, if no relief is had, to insanity.

Travellers describe the punishment inflicted in China on criminals sentenced to be kept awake till they die, as the most terrible punishment ever devised by the diabolical ingenuity of man, for tormenting a fellow-being. The closing scenes of the shocking condition to which the unhappy prisoner is reduced, are painful in the extreme. He finally becomes insensible to almost every form of torture that can be inflicted to keep him awake, and dies at last, about the fifteenth day, in awful misery.

Two criminals in Russia, not many years since, were made the subjects of a scientific experiment in regard to the value of sleep in the maintenance of life. They were kept awake with the utmost difficulty, after eighty hours. What fiendish cruelties were practised on the wretched creatures beyond wedging their heads, so as to be continually receiving droppings of cold water, has not been revealed; but on the nineteenth day, death mercifully terminated their misery. Such punishment is a disgrace to any country, and too shocking to be tolerated where Christianity is the religion of the rulers.

### CONSTITUTIONAL STAMINA.

A sound constitution must have its beginning in childhood. Small girls, anywhere from three years to ten, should sleep in good-sized airy rooms. It is not always possible or convenient to provide them spacious apartments, but it is possible to ventilate their dormitories thoroughly daily, in a house with doors and windows. On a free introduction of air, vital elasticity and recuperation of the sleeper mainly depends. Of the value of an uncontaminated atmosphere, no one entertains a doubt; therefore, the discussion of a subject so frequently before the public as ventilation, is passed over in silence, its importance being understood, and everywhere appreciated.

When too many persons occupy the same apartment, even if of large dimensions, the vitality of the air is ultimately diminished very considerably, which is recognized by an increased temperature, perspiration, and physical exhaustion. Small rooms, in the occupancy of two persons, are soon in a similar state, if no fresh supply is regularly admitted.

Two children sleeping in separate beds thrive better than when together in the same bed, even in a spacious room, high studded, and in other respects appropriate, because they are kept from inhaling each other's breath, hardly to be avoided in their unconscious relations in sound sleep.

Expired air is charged with elements deleterious to other lungs, and especially so if from a person indisposed or sick. Expired air directly from the mouth or nostrils is deprived of all its vital properties. If inhaled into the lungs of another, it is particularly injurious. No doubt, many painful forms of sickness in children, which cannot be accounted for on familiar principles, have an origin in the baneful inhalation of another's breath.

A lady exposed to incidental inhalations of the offensive breath of a smoking husband, or one whose expirations are laden with alcoholic odors, is liable to various forms of indisposition, the result of Nature's efforts to drive out of her system the cause of disturbance.

Expired air is deprived of oxygen, the pabulum of life, while carbonic acid, destructive to life in its highest forms, constitutes the volume of breath thus expelled from the lungs, mixed with aqueous vapor and impurities, which chemistry detects. Other vile products, traceable to tobacco and whiskey, are also carried off in the breath.

When such expired compositions are drawn into the sound chest of a sleeping companion, although only occasionally, an incalculable amount of future suffering may be thus unsuspectedly commenced, which medical skill cannot always successfully control.

Growth, strength, and the regularity of organic functions, perfect nutrition and mental development, are each and all of them defective, if the air is charged with deleterious elements, or simply deprived of oxygen.

Each one is entitled to as much pure air as their organization requires,—the lungs being the instruments for separating the constituents of which it is formed, and conveying such elements into the circulation as support life, and rejecting those which are noxious.

### TRANSFERENCE OF VITALITY.

A pale, feeble, sickly appearance of children, whose debility cannot be clearly accounted for, and made the more mysterious from having a sound healthy parentage, not unfrequently are amply provided with all the appliances for their comfort with one single exception,—their sleeping-room.

It is a wise precaution, therefore, to place girls in separate beds, and better still, give each one a room exclusively to herself. Neither is it proper for sufficient reasons that might be given, for children of different ages to sleep in the same bed, even when ventilation and the dimensions of the apartment are satisfactory.

When two children are thus associated for eight or ten hours, it has been ascertained that, if either becomes indisposed, it is usually the youngest, although both were in the beginning equally well and robust.

Physicians recognize a law of which very little is known beyond the effects resulting from imprudence, in placing persons of different ages under circumstances which lead to an actual transference of vitality from one to the other, at the expense of the one from whom it is drawn.

By placing a strong and a feeble child in bed together, after a few months the latter will profit physically, while the other will lose some of its former freshness and vigor. If a sound, plump, healthy child sleep with an emaciated, sickly, or aged person, the former becomes indisposed. Therefore, children of a tender age should not be the bed companions of aged aunts or grandmothers.

Sometimes a blooming child is unaccountably reduced in strength, loses its rosy cheeks, and moves about languidly, losing its relish for food,—which may result from sleeping with an aged person.

A feeble, attenuated woman, advanced in years, will wonderfully recruit by sleeping with a healthy child. She mysteriously imbibes vital force from the innocent in her withered arms.

How that subtle something that passes from one to the other is transferred, or what it is,—has not yet been philosophically demonstrated. The fact, however, that some property does escape from one, and is taken up by the other, is not questioned by medical men.

It is not judicious, therefore, to have a nurse who has passed beyond the middle age of life, for an infant. She will take from the child, by this law of transference, more than the child will receive from her. It is equally unsafe to place children of a tender age in sleeping-rooms, or in bed with servants or nurses who are ten and fifteen years older, or have sallow complexions, decayed teeth, a bad breath, or peculiar habits of any kind.

These precautions have express reference to young female children. But it would be equally injudicious to permit an athlethic, energetic boy to be the habitual bed-fellow of his grandfather.

Such violations of the general laws of health are not so common in regard to boys as girls. Aged women are particularly fond of sleeping with their young kindred. Their sympathies are active, and their love for the society of children rather increases than diminishes with the progress of years.

The extraction of vitality was far better understood by the Jews at an early historical period, than by modern teachers of hygienic laws, with all the assistance and appliances of modern discoveries.

When King David was waning in health, and the alarm spreading that his life was in danger, on the philosophical principle recognized in this chapter, effort was made in his behalf to transfuse vitality into the monarch's cold and fragile body, by taking it from a very select source. But the hopeful experiment was deferred too long. He could gather no warmth, in the language of the sacred narrative, and the king gave up the ghost. The theory was correct, but it was put in practice too late to be of service.

When a young man, for the worldly consideration of property, weds a woman old enough to be his mother, she will gain by the contract in health. Repeated instances of ill-assorted marriages of that description have established the fact, that the husband will decline in health, with all his advantages of youth, and generally die first. His vitality is transferred.

This may be novel intelligence to those who are more intent upon bettering their financial circumstances by matrimony, than in securing happiness in that sacred relation.

In those reprehensible and unnatural matches, where selfishness is the ruling passion, an aged wife, in a majority of cases, will become a widow.

Reversing the proposition, the husband being the oldest by years enough to have been the father or grandfather of his wife, although so much her senior, may outlive his young wife.

There are many deviations from the principles laid down in these observations, but individual cases do not conflict at all with this peculiar law in reference to the transference of vitality.

When a young woman sells herself to a man old enough to be her grandfather, she puts her life in jeopardy. She usually dies first. There are modifying circumstances, sometimes, that partially arrest the downward tendency to a premature dissolution, of which the public are ignorant. Family secrets embody physiological problems more strange than poetic fiction. Of the many who thus run the gauntlet for luck in marital adventure, a few win the race, living to get what they anticipated—wealth. When women have attained it by a sacrifice, they deliberately survey the ground, and select a second husband more congenial to their age, to fill an hiatus in their affections.

It is a fearful risk to marry a husband considerably the oldest. There should be a correspondence in age, as in temperament and· disposition, to secure all that a divine institution promises to those who are guided by reason, rather than impulse, on entering upon the solemn obligations of matrimony.

## LAW OF ADAPTATION.

A man should not be much more than nine years older than his wife. From four to seven years the senior is a natural relation, and always insures a reasonable prospect of domestic happiness. Their physical, intellectual, and moral natures then harmonize most satisfactorily.

Leaving out ambitious views in regard to .advantageous alliances, from a selfish determination to sacrifice yearnings of the heart for ,pecuniary power, if the husband is a few years older than his wife, both parties will have more domestic comfort than when madam is the senior.

In regard to sleep, as especially belonging to the domain of health,—it may be received without qualification, as both sound and reasonable, that two women accustomed to sleep together would escape many annoyances in the form of headaches, neuralgic twinges, occasional nausea, etc., were they in separate beds.

It is injurious for two men of about equal age to lodge habitually in the same bed, but always worse for females. Young women, at all times after the establishment of perfect womanhood, should lodge alone. The objections to sleeping together are not removed, even though the apartment is large and airy.

Husbands and wives sleeping in the same bed do not contaminate the air, as two men or two women do. There is a correcting influence from opposite sexes thus circumstanced, difficult to explain, but, nevertheless, true. In many parts of Italy they practise the discreet policy of ·never permitting two persons to occupy one bed, by making them too narrow for two. It impresses the traveller with curious surprise to see hotel beds in that sunny land so very insignificant in width.

There is a peculiar electrical condition of the sexes. Two females do not develop the same nervous state, neither is it produced by two men, that is, elicited by one of each sex. The extreme subtlety of this phenomenon defies scrutiny. We really do not know anything about it beyond the fact, which is familiar knowledge with those who have no insight into the first principles of science.

It is said that a man and a woman, introduced into a perfectly dark room, totally ignorant of the presence of each other, will not only soon ascertain that a person is present, and that without moving an inch, but decide accurately whether the neighbor, unseen and unheard, is a man or a woman !

## FEMALE OPERATIVES.

One reason why female operatives in large manufacturing establishments, as cotton-mills, book-binderies, printing-offices, paper-box shops, tailoring lofts, etc., are pale, cadaverous, or yellowish, besides being of inferior strength, although but a few months thus circumstanced, is due to exhalations from their own bodies, inhaled with the air they are breathing.

A morbid craving for clay, charcoal, slate-pencils, chalk, broken bits of crockery, and similar substances, is almost irrepressible among females when working together in considerable numbers. This is usually regarded as a novel circumstance.

Deprived of home influences, grouped together in a vitiated atmosphere, morbid propensities are generated.

Such was the charcoal-eating propensity of the female weavers in one of the great mills in Massachusetts a few years since, orders were given to lock the bins in which charcoal was kept, as the girls were actually consuming such quantities daily.

## TEMPERATURE OF THE BODY.

In drawing-rooms, halls, concerts, and, indeed, on all public occasions, where large numbers of persons are compactly wedged together, ladies, much sooner than men, complain of a sense of suffocation. While gentlemen are quite at their ease, the feminine part of the audience are plying fans with extreme activity.

In churches, men sit bundled in thick heavy clothing, buttoned to their chins, and then are only just comfortable, while ladies throw off their outer garments, and express by various movements their oppression from heat or foul air.

In public conveyances, nothing is more common than to have a car full of men thrown out of temper by the entrance of a frail, shadowy woman, who immediately requests a window to be opened. On some railroads, cars are expressly appropriated for females, in which they may have a temperature as much below zero as their necessities require; but they invariably indicate dissatisfaction in being placed by themselves, even though they might respire more agreeably.

Clothing which women wear is more delicate in texture, thinner and lighter than male attire in the same climate. Yet they are as warm and comfortable as muscular men in their McIntoshes and buffalo overcoats.

This shows that women have a temperature above the sluggish vitality of their legal protectors. Their circulation is more rapid up to about forty-five, *ceteris paribus*, than the circulation of men, sustained upon the same diet, and having a home in common.

## SLEEPING WITH ANIMALS.

The importance of having women sleep well—that is, refreshingly—need not be argued. A vile practice is gaining in this country, that should be frowned down by all well-wishers to humanity. Young ladies, and particularly many in the maturity of age, are excessively fond of pet dogs. They are their most intimate companions, and they bestow as much attention upon them as affectionate mothers mete out to their children, to gratify their philoprogenitiveness. It must be met by something, and black-and-tan imps take the place which poor, abandoned orphans should have in their arms and in their affections.

They not only feed them on delicacies unsuitable to their natures, but they take them out to ride in carriages, when it would please them more to have liberty to run on the ground, like all quadrupeds. It is disgusting to see little snapping curs receiving the fondest caresses and the sweetest tones of endearment, lavished on them by accomplished women who would not allow imploring poverty to stand between their ladyships and a darling puppy.

There are demoralizations and contaminating influences connected with this canine mania, which a loving father is bound to forbid. If his commands are not honored, his next resort should be a revolver, which would most effectually rid the premises of such unnatural and such disgusting associates of his daughters or his wife.

Not satisfied with feeding their dogs with dainties unsuitable to their organs of digestion, their carnivorous maws are filled with such articles as they like best themselves; they pamper them on cushions, walk with them reposing on their bosoms, and sleep with them ostensibly at their feet. The rage for pet dogs is a cultivated taste. They commence with moderate attentions, but soon become fascinated, and next bewitched. From a pillow on a rug, they are promoted to the foot of the bed. Having served a sophomorical period there, the rise to the position of senior and intimate companion is not distant.

"Whoso lies down with dogs will rise up with fleas," says the proverb. It cannot be healthy for a woman to inhale air a dog has breathed, to say nothing of the emanations from the pores of his body in the confined apartments in which such favorites are ordinarily kept.

There is a tremendous exposure to an incurable malady, if, by any mishap, madam or her daughter should be bitten by a rabid pet. They become mad, and no dog is proof against a sudden development of that incurable malady,—hydrophobia.

Cats are preferable to dogs for little children, in their kitten-hood days, as less prone to bite and snap, even if handled roughly.*

---

* June 8th, 1871, the following circumstance occurred in a police court in the city of New York, which shows how strong an attachment for a dog may become:—

Mrs. Sophia Clinton lived at 156 Clinton street. She had a little black-and-tan dog, and the black-and tan dog's name was Dexter. A week ago the dog strayed away or was stolen, and she advertised in the papers, and searched the metropolis for that little dog. At last she found him in the possession of a German named Lippman Kessler, living at 130 Attorney street. But Mr. Kessler would not give up the animal. So Mrs. Clinton had Mr. Lippman Kessler arrested, and he was brought before Judge Scott, at Essex Market. Quite a scene ensued as the high disputing parties made their entrance into the vestibule of Justice. Mrs. Clinton is a tall, slender lady, of fine presence, and has beautiful blonde hair. Mr. Kessler is a gross-looking Teuton of herculean build. The lady was very demonstrative in her affections, and kissed and hugged the "innocent cause of the war," calling him "mother's own baby," and other endearing terms. Poor little "Dexter was lost, wasn't he? Poor little pet!"

## CRIBS FOR INFANTS.

When a child has been weaned, it should have a crib by itself. With the development of teeth, it is a sign a modified aliment is required, and their food should have more solidity. No rules can be given, nor are they required, for feeding young children. No arbitrary system of dieting can be borne. Variety is necessary, that elements may be selected by fashioning vessels essential in their economy.

If a child of a tender age is habitually fed on diluted milk, softened biscuit, rice, tapioca, and similar unsatisfactory pap, because an ignorant mother has a theory which becomes a law in her own house,—if it lives, it can hardly escape having a defective mind encased in a feeble body.

## OVER-DOING.

Thousands of children die annually that would have lived, had they been let alone. One of the trials of infancy is teething. Large numbers are chronicled in the bills of mortality as

---

"Oh, yais; zay Dechster! Dechster! mooch vot you bleese. I call heem Preence. He coom shoosd de zame," said Mr. Kessler.

"What mark do you know him by?" asked the judge.

Mrs. Clinton—"His claws were cut short, so he would walk nice, and his ears are cut longer than most dogs'; and, Judge, here is the man that cut his ears;" pointing to a young gentleman standing alongside.

Mr. Kessler—"Oh, yais. You hear owel aboud dem tings fon de boleeceman. Coom here, Preence, coom. You see, Shudge, he coom to me yust de same;" and the little dog trotted over to his last owner.

Judge—"Where did you get the dog?"

Kessler—"I got heem fon a shoemaker man. I dond can remember his name, dere is so mooch excitements about dot."

Mrs. Clinton called the dog back again, and it clung to her, as if it had regained its mistress. At last the Judge decided in her favor, and she stalked off triumphantly.

having died from that cause. The truth is, if the facts could be known, children are doctored to death far oftener than they die from diseases peculiar to their age. Indifferent physicians guess at their ailments, and prescribe accordingly, without much reflection, since to do nothing, when called in for advice, would be rather unprofessional.

Charging the stomachs of little children, who cannot give any account of their indisposition, with nauseous drugs, is reprehensible. More vital effort is wasted to throw them off, than would have been expended in resisting the invasion of inflammation of the gums in the protrusion of primary teeth.

Let infants and young misses have separate beds. School girls should invariably sleep by themselves. When they become young ladies, it is inexcusable to permit two of them, however attached and dear to each other as friends, to occupy the same beds habitually.

Pulmonary consumption is sadly sweeping away women from spheres they beautify and adorn. The mortality is far beyond what it would be from hereditary sources, because those who die of it transgress many laws of health. To obviate the formation of a susceptibility in the constitution to the approach of pulmonary consumption, begin seasonably by simply avoiding exposures to influences which may be derived from sleeping with others in early life.

## BEDS.

There is another subject connected with this topic, too long overlooked, which it is proper to introduce. The materials of which beds should be made is an important study.

It is certain that there is a constant exhalation from the surface of the body. If the emunctories are closed by inflammation, or accumulations of foreign matter, a thickening of the epidermic tissues, indeed, from any source, as exanthematous obstructions, produces internal febrile heat and universal disturbance in the system.

Febrile heat sometimes ensues on mechanical principles,—from the non-escapement of fluids which ought necessarily to pass off externally. Insensible perspiration is a safety-valve for the body, as much as a crater of a volcano is the natural outlet of pent-up forces that would destroy the whole mountain if not allowed to escape.

The kidneys by no means secrete all the fluid taken into the stomach in a very warm day.

Fluids taken by the renal apparatus directly to the bladder, hold in solution elements already referred to as being commingled with our food, but hurtful if not carried off in the most direct manner.

Those who perspire freely, when exposed to a slightly elevated temperature, have thus less duty imposed on the kidneys.

### OFFENSIVE CUTANEOUS EXHALATIONS.

Persons who perspire easily, and more than others under ordinary circumstances, rarely have either dropsy or renal difficulties.

There is a singular difference in the character of cutaneous transpirations in different persons, detected by the sense of smell, but not by the individual from whom it escapes. It is offensively unpleasant to the olfactories from most colored persons, particularly when they have been exercising or in warm weather.

That disagreeable odor is not without its use in the general economy of things. Africa abounds with annoying insects, the

torment of humanity, as of all animals. So particularly offensive is the perspiration from the bodies of the natives, it pro-
tects them, like an invisible cloud, against attacks of swarms
of pestiferous flies, gnats, and winged plagues of indescribable
forms, which no life could resist, were it not for that curious
provision for defense.

Apiarians are familiar with what every body know, that
bees cannot tolerate the presence of some persons, while others
may handle their hives, extract sheets of comb, or swarm new
colonies with perfect impunity.

This is accounted for by the ignorant, on the presumption
that honey-bees recognize an enemy in the one or a friend in
the other. No doubt those who annoy them by their presence
to exasperation, give off an offensive vapor which the acute organization of the bees detects as a nuisance. Those who fearlessly explore the interior of a hive, and even suffer bees to
light upon them without being stung, exhale no vapor that
meets their disapprobation.

This is, probably, the whole secret and explanation of the
supposed friendship or hostility of honey-bees. The perspiration of intemperate persons, as well as those excessively given
to the consumption of tobacco, is laden, unknown to themselves, with exceedingly offensive matter, which is quite as
disgusting to those brought within the sphere of its emanation,
as to the quick discrimination of honey-bees and wasps.

### PROGRESSIVE DECOMPOSITION IN LIFE.

There is a constant, uninterrupted process of decomposition
going on in every organ and tissue of the body of every living
being. When a new particle is placed in position, an old one
is removed.

There are but three ways of throwing off effete, dead matter, viz., through the pores of the skin, the intestinal tube, and the bladder. To do this, the blood holds immense amounts of *débris* in solution. When long retained, physicians speak of a bad condition of the blood. Quacks, without knowing anything about it, harp incessantly on its impurity, and get rich on the sale of nostrums for its purification.

Such medications are absurdities. It is ridiculous nonsense to prate, as these irresponsible speculators in health do, about pretending to physic the blood. It is as impossible to produce any such operation as it would be to bombard the sun. Charged, as that vital fluid must be always, with worn-out materials, which have served a purpose till all of value in them had been exhausted,—it is a natural process to be floated away, and nature will take care of herself without the aid of pseudo-medical specialists.

Tonics, properly directed, may assist a debilitated invalid by giving vigor to some flagging organ, in this never-ceasing process of receiving, appropriating, and then setting at liberty that which ceases to be any longer of utility.

Avoid one probable cause of indisposition, as far as possible, by breathing good air rather than foul, if just as readily obtained.

Feather beds yield, in the atmosphere of a close room, a peculiar mephitic odor, traceable to a slow decomposition of the tubes of the feathers. Years are required, if no artificial efforts are made by severe kiln-drying or baking, before feathers lose that offensive character. Even after various expedients for airing them by drying, they re-imbibe moisture, and the old odor is again given off.

Thus, the best directed efforts in purifying feathers are only temporary, and, therefore, they should be abandoned. It

is just as injurious to be inhaling every night the impure air
of a room in which a feather-bed putrefaction is progressing,
as to have the decaying carcase of a dead animal under the
bed, from which sulphuretted hydrogen gas was escaping.

In northern climates, where the progress of feather decomposition is slowly conducted, feather beds are common, and
less to be dreaded than where the summers are long and hot.
But they ought to be given up wholly and entirely, as they
probably will be, with a more general diffusion of the principles of hygiene, the importance of which is happily
beginning to be understood.

Even when the weather is cold, the heat of the body
actually penetrates to the feathers, acting chemically in setting
free an unpleasant odor, if the room is not well aired. Under
any circumstances, those of delicate organizations, subjected
to severe exposures which affect the lungs, should avoid
feather beds. So should asthmatic people. Emanations from
a feather pillow, even when the bed is of hair, or some other
common material, will sometimes bring on a stricture of the
bronchial tubes, so severe that the sufferer can scarcely draw
in sufficient breath for sustaining life. Asthmatics should
shun feathers in beds, bolsters, or pillows.

Wool beds are admirable. They are warm, soft, and elastic.
They have been objected to on account of being an animal
product, as well as feathers. But, admitting that decomposition must, of course, be the destiny of all animal matter, in
whatever form it may be utilized, there is really no such
cogent reason for rejecting wool as feathers. We like them,
and recommend them for invalids of a spare habit.

Next, hair-mattresses, in universal use, while fresh and
new, are delightful beds. But they are an animal product also,
very likely to be preyed upon by minute insects which cut

the hairs into bits much sooner than suspected. An old hair-mattress is a living sack of abominations, in which life, death, and successive generations of mites, too minute to be seen without a magnifier, undoubtedly give rise to eruptions, cutaneous irritations, and perhaps unpleasant conditions of the mucous membrane of the lungs, from breathing air laden with matter escaping through the tick.

Many a traveller has imbibed the seeds of death by sleeping on such kinds of beds in hotels. They would be gainers by sleeping on the floor, rather than recline on an old hair-mattress which may have been soaked with the offensive sweat products of a sick stranger the night before, or be in a state of slow chemical putridity, from which gases are given off that may generate disease which no medications could arrest.

Frequent opening of the sacks, repicking and drying in open, brilliant sunlight, and thoroughly drying beds of all kinds in hotels and boarding-houses, should be enforced under police inspection, as a measure for securing public health among other sanitary precautions so well received by the public.

Cotton-wool beds have been introduced, but not very successfully. They mat and become extremely hard, soon losing all the elasticity they may have had at first. Besides, they imbibe moisture which is difficult to expel in such a thick mass.

Within a few years, sponge beds have been introduced, which have their friends, especially among those interested in the sale of them. There has hardly been time to ascertain their true merits. If their elasticity, when chopped or torn into fragments, depends on being made supple with glycerine, by and by objections will be raised against them. However, they are not to be criticised unfavorably till more is known of the advantages they present.

One of the latest and best yet presented for acceptance, is the metallic. In appearance it is a wire tick, woven, or made of rings linked together, fastened by its edges to the inner margins of the bedstead.

They are always clean and free from collections which attach to other beds. Being galvanized, they neither rust nor become dark-colored. Water beds, which were thought particularly valuable for hospitals, have not been in general use. The metallic bed addresses itself to the commonsense of a very limited intelligence as valuable. A mattress is rarely required on them. A few thicknesses of soft woolen blankets are quite sufficient; they are soft and yielding to the form of the sleeper. In a word, they are admirable and appear destined to be extensively adopted wherever large numbers of beds are required in any one place—as on shipboard, hospitals, barracks, and hotels. Families ought to give them a decided preference.

There is immense economy in them. Beside all the properties found in other beds, of giving ease and comfort, they present none of the objections cited in reference to feathers, hair, wool, cotton, rattan, husks, or straw. No insects will ever burrow upon them;—and when injured or broken, or they become valueless for the purposes for which they were made, they may then be sold for old iron.

In fitting up a private dwelling, the economy of the iron bed is apparent. They are the least objectionable; and the very best for young persons, especially children, because they would be perfectly free from moisture and vermin. They can be set into any kind of bedstead, wood or iron, but iron should take the place of wooden bedsteads. It is the bed for women— incomparatively superior to any other kind in use.

# CHAPTER XIII.

## THE FOOD OF WOMEN.

Dietetics of the World—Everything Eaten—Difference of Taste—Habit—
Sugar a Necessity—Economy of the Liver—Pork—By whom Avoided—
Starch—Experiment with Honey Bees—Law of Life Illustrated—Fruits
to be freely given to Children—Open-Air Exercise for Girls—A Benevolent Citizen of Boston—Fish Excellent Food—And Why?

EVERY creeping thing, even disgusting insects, vermin, reptiles, lizards, and crawling ophidians, are used for human food.
Of course, they are not appropriated for that purpose in civilized countries; but, with savages and barbarians, whatever will
sustain life is greedily seized upon without reference to external
appearances, habits, character, or flavor. Necessity compels
many tribes to sustain themselves on food that would not have
been selected from choice or a depraved taste, if anything else
could be procured.

Under such circumstances, it has not been discovered that
those who feed thus promiscuously and offensively, measured
by the standard of civilization, are any more prone to sickness,
or are shorter-lived, than gentlemen and ladies who dine
sumptuously on roast beef and pudding.

"Slay and eat," was a command to Peter, when the sheet
was let down before his eyes, filled with all manner of strange
forms. It is a maxim in law that circumstances alter cases. So
it is in respect to diet. Our impressions respecting the whole-
someness or unwholesomeness of particular kinds of food, are
formed from the remarks, or likes and dislikes of those with
whom, and among whom, our early associations were established.

We are influenced, without being able to explain why, by what others say and are practising within the circle in which we are moving. Our social education, which is entirely independent of letters, books, or schools, is commenced and completed very early in the family. What we learn there abides with us ever after. We cannot emancipate ourselves from the errors thus imbibed, nor free ourselves from the cordon of responsibilities with which we feel ourselves surrounded, without violating moral laws on which both safety and happiness seem to depend.

Nearly all we know upon the subject of food comes from the experience of others, and rarely from our own. Had we been accustomed to swallows'-nest soups, a rich dish in China, we certainly should have had no prejudices to contend with in later years, were it served to us; but, never having tasted it, the very thought of such a singular preparation for the stomach is nauseating. Precisely so, also, in regard to that still more disgusting delicacy of almond-eyed races, *beche de la mer*, a large, slimy, soft, hideous-looking sea-slug, held in the highest estimation in aristocratic society throughout the whole of China.

Sailors on a wreck have fed upon the decaying corpse of a starved companion, without any of those painful results which theoretically follow from eating putrescent food. Hungry Bedouins feast upon dried locusts; roaming savages of Africa satisfy a voracious appetite with a roasted boa-constrictor, or a baked monkey. A Mexican slaughters a cow for the sake of a dainty morsel, the half-grown calf, throwing away the beef of the mother, as too coarse and too common for a refined and cultivated gourmand. Stewed puppies were a choice preparation when Captain Cook discovered the Sandwich Islands. Fried eels, boiled snails, five-fingered Jacks, oysters, prawns, clams, shrimps, etc., which belong to our catalogue of modern eatables, are quite as objectionable, contemplated as awful-looking creatures, as many things we exclude from gustatory favor, on account of their imagined bad qualities.

## PUTRIDITY.

If chemical decomposition is not so far advanced as to destroy cohesion, no unfavorable effects result from eating anything that has once been alive. Those animals which have sacks of poison in them are excluded, as well as those that secrete an abominable fluid from particular glands, which, in both cases, are defences against their enemies.

Pampered city gourmands keep venison till it becomes partially decayed, before it attains that delicious flavor which meets the approval of an aldermanic stomach.

To those unaccustomed to that delicacy—a conversational theme of officials, dieted at the expense of taxpayers—such a meal would seem freighted with death in the pot, especially when a smoking quarter comes to the table, a perfect nuisance to uneducated olfactories.

Overcome the pangs of hunger with whatever is of an animal origin, there are properties in it which a stomach fashions to meet the exigencies of the system. No carrion is too corrupt for some carnivorous beasts and birds. Contact with the gastric juice deprives it quickly of the taint, while the decaying softness of flesh is thus prepared for rapid digestion.

Man has always been, and always will continue, to sustain himself on a mixed aliment.

## SALT.

The late Mr. Sylvester Graham was a prominent vegetarian

fanatic.  He was even extremely prejudiced against salt.  He could not abide it, and exerted his vocal skill in trying to convince silly old women, of both sexes, that eating salt was about equal to taking in death by grains.

How supremely ridiculous to be at war with the law of necessity!  There is not a treatise extant on health, held in esteem by competent scientific authorities, that does not admit, unequivocally, that common salt is found in all our tissues and fluids.

We could not be what we are, or what we may be, without it.  Salt is found in almost every article made use of as food, whether the newly-fledged school of ignorant physiological reformers approve of it or not.  By the introduction of salt into the system, the blood globules are supposed to be sustained in their form, and prepared for the purposes of life.  Even the tears we shed contain salt.  We must be supplied with it.  Nature, therefore, in anticipation of the necessities of all warm-blooded animals, was careful to introduce it into vegetables, and from them the flesh and fluids are kept supplied.

There are conditions in which the supply from that source is not equal to the demands of the body.  Where the quantity secured by plants and grasses in some latitudes is too small, the deficit is met for man by commerce.  Buffaloes, deer, etc., in the primitive state of this country, came in droves from great distances in the far West towards the Atlantic where salt springs abounded, to obtain what instinct compelled them to seek; or suffer and die, if not found.

Grass-feeding animals search for it in their wild state continually.  Whenever they discover a saline quality in water, that spot is not only remembered, but intelligence of its locality is extensively propagated and transmitted from one generation to another. How that was accomplished without articulative language'will ever remain as much of a paradox as the propagation of the intuition of birds to go South for winter quarters which is understood by the youngest of the flock for the first time leaving the neighborhood where they were reared, for a flight of one or two thousand miles to an unknown region.

The health of wild or domesticated animals imperiously demands salt. Some are so organized that what they obtain in their food is sufficient for them. The farmer feeds it out at stated periods to his stock. Timid horses may be caught with a handful, when nothing else would tempt them to yield up their liberty to become the slaves of their captors.

Carnivorous animals, flesh-eaters exclusively, obtain in their prey just enough of the saline element to answer the physical needs of their organization.

## ANOMALIES.

No two persons are constituted so nearly alike as to perfectly agree in their taste or appetency for food. One may object to pastry, while another loves it dearly. A small amount of meat suffices for some, others have no relish for it at all. Vegetables are coveted exclusively by some individuals. In them are provided sugar, starch, gelatine, etc., required in the reparation of their tissues. Thus, bread, in universal request, contains some, if not all, of those elements, and, therefore, each sustains himself on an article in which some, if not all, the life-sustaining properties exist, necessary for his preservation.

Bread, by baking, is prepared for being converted into glucose, soon after reaching the stomach. It is changed into a sweetish paste by a vital chemical action.

Sugar is indispensable. If the supply is too small from without, the liver, as set forth in another chapter, immediately manufactures enough to supply the deficit necessary in the economy of the individual.

## AFFECTING THE TEETH.

A common opinion prevails that sugar is injurious to the teeth. A grosser mistake was never propagated. Carious teeth, denuded of enamel, ache when sweets are in contact with the decaying surface; but the cause of the caries is due to other agencies, and not to sugar.

Children crave it, and the universal desire for sweets gives employment to immense numbers of laborers in tropical countries to meet the demands of those parts of the globe where it cannot be made.

Wherever civilization has raised its standard, sugar becomes a staple commodity. How preposterous, then, to attempt turning back the current of trade, or interfering with the great movements of commercial activity, because, forsooth, some addle-headed theorist wishes to immortalize himself by opposing constitutional tendencies of improved and improving humanity.

## SUGAR HAS IGNORANT ENEMIES.

Opposing the use of sugar and salt is simply to expose one's imbecility, want of judgment, and limited views of nature's unalterable laws.

## OMNIVOROUS.

Teeth, stomach, and their auxiliary appendages are constructed upon principles of relationship to secure perfect nutrition. Because men can subsist on a mixed diet, is found their ability for traversing the globe from the tropics to the frozen regions of the poles.

We go with impunity from arctic ice-fields to the burning sands of an African desert, through all extremes of climate, without apprehension of not being able to sustain ourselves on any kind of food which may be offered. No animal could be transported through such diversified climates, and feed on diversified products as they might present, without perishing. They must have the element especially fitted to their organization. If that is not to be had, they perish.

Swine, and some birds, to a limited extent, are omnivorous; still they cannot thrive when removed from their natural habitat, unless provided with food analogous to that in which they attain their highest development.

Feathered tribes feed largely on insects, larvæ, seeds, etc., which is a mixture of animal and vegetable food. Were it otherwise, in their periodical migrations sad consequences would follow. An omnivorous appetite can be accommodated in different localities, where animal and vegetable products abound, without impairing muscular force, or unfitting them for returning to such food as they subsist upon a part of the season North or South.

## PORK.

Swine feed indifferently on flesh, vegetables, or garbage, without reference to its composition, even in a state of putrefactive fermentation. A knowledge of their habits may have led to a prohibition of pork under the Mosaic dispensation. They have always been held as unclean, and, therefore, unsuitable for human food, from a remote antiquity, by both Jews and Mahometans.

Shaker communities in this country have uniformly abstained from pork. Trichinus spiralis, which affects men, women, and children, is traced directly to swine. Their minute eggs taken into the human stomach, or, indeed, any stomach, as far as we know to the contrary, resist the gastric juice; although it dissolves metals, their vitality resists its potency. Trichinæ reach the muscles, and the tapeworm keeps possession of the alimentary canal.

Scrofula, which is an enlargement and tumefaction of the glands, is also believed to be aggravated, if not produced by pork. The term scrofula is derivtaed from a word indicating filthiness.

Shakers are remarkable for their fair skins, clear complexion, and exemption from scrofulous affections. They very rarely have either cutaneous blotches, discolorations, moles, or eruptions, which confirms them in the opinion, that they are right in excluding pork from their tables, and living swine from their farms.

When swine are fed, as they usually are in the vicinity of cities, and populous towns, on offal gathered in carts, the backyard accumulations from kitchens, sour, decomposing, or offensive, no pains ever being taken to preserve such collections from passing into a stage wholly unfit for food, their flesh becomes diseased, in consequence of being compelled to subsist on a vile hodge-podge—a perfect salmagundi of concentrated vileness.

Raised, as they are in back settlements of the West, on mast, which they gather in their free rambles in the woods, or when stall-fed on sound corn, the pork is less objectionable, and not likely to be diseased.  But it is safer never to use it, since it is difficult to decide in the market from whence it came, upon the quality.

Ladies who are anxious to preserve their fair faces free from roughness, redness, eruptive pimples, and glandular enlargements about the neck, must shun pork.  They cannot breakfast on sausages without running a greater risk than with a pork-steak; because they are usually made of scrapings of bones, or the poorest quality of pork, so compounded with pepper, lard, and pulverized herbs, as to conceal the objectionable appearance or taint they would give out, were it not for salt, and the deceptive skill of the manufacturer.

Smoked hams pass through processes which are thought to destroy parasites burrowing in the best of them.  By severe boiling or baking, minute eggs deposited in them are effectually destroyed, so that in that form, if pork is at all allowable, it is in a thoroughly cured and thoroughly cooked ham.  Even when a long while smoked, if taken in sandwiches, raw—sometimes practised—there is undoubted danger of being infected with trichinæ and tape-worm eggs.

### DISCOLORATION OF THE SKIN.

Moth-spots, those irregular yellow patches that appear on the chin, side of the nose, on the forehead, and near the ears of middle-aged women; irritable eruptions on the limbs, known as salt rheum; excess of dandruff on the head; moles, and spongy outgrowths,—are each and all of them aggravated by pork.

Cooking exceedingly modifies food for being more easily and rapidly assimilated.  Hard and unpalatable articles, in a raw state, are quite savory when subjected to culinary operations. But of all laboratories, the stomach is the most perfect. Vital chemistry is superior to art, and whatever enters the stomach is subjected both to mechanical and chemical influences, before the absorbents draw upon the mass for nutriment.

### FURTHER OBSERVATIONS ON SUGAR.

Sugar is an ingredient of most plants, roots, fruits, and grains on which animals subsist. So acute is the sense of smell in quadrupeds, especially the wild ones—and always active in the domesticated, as horses, oxen, deer, sheep, and goats—they select, with extreme care and accuracy, only such vegetables as yield it.

It is for our interest, as it is for the promotion of individual and public health, to cultivate plants and roots for our domesticated animals that contain the largest per cent of saccharine matter. Hence, beets, carrots, and turnips are excellent for them.

We have nothing to do with commercial interests in these deliberations. Reference is simply made to those products that have an influence on individual health.

Next to sugar, in the order of dietetic indispensables, are the cereals.

Starch passes through several interesting stages before it yields those elements on which its nutritive properties depend. Flour is first made into dough, and by baking is changed so materially, as to be wholly unlike its appearance either in flour or dough. Both in sapidity and in quality for the support of life, the processes through which it passes from the mill to leaving the oven, are remarkably curious; yet so common and familiar, a thought is never bestowed upon the subject except by teachers or writers on the phenomena of digestion.

If compelled to subsist upon any one article exclusively, even if it contains sugar, it ceases to be serviceable to man, but most of the animals live in excellent health through their whole allotted lifetime, as regulated by the law of limitation, on one kind of food.

Our food must be frequently changed, or compounded with different ingredients. Dogs, cats, and, indeed, all the carnivorous animals, are quite independent of sugar from plants or fruits, all that they require being manufactured within their own bodies. If they are fed on sugar a little time the relish for it soon subsides, and they lose flesh, become feeble, and die.

The following table shows the amount of sugar in fruits and grains, with which we are most familiar. Nature has made ample provision for the necessities of those whose organization requires it.

In one hundred parts, sugar is in the following proportion:

| | |
|---|---|
| Figs....................62.50 | Wheat Flour........5.20 to 48 |
| Cherries................18.12 | Rye Meal...........3.28 |
| Peaches................16.45 | Indian Meal........1.45 |
| Pears...................11.52 | Peas................2.00 |
| Tamarinds..............12.50 | Cow's Milk......... 4.77 |
| Beets.................... 9.00 | Human Milk........5.50 |
| Barley........ .......... 5.21 | |

Fruits abound more in sugar than grain, but the latter furnishes starch. We cannot subsist on either alone, so well or so long as when compounded with other materials.

There is an inborn love for sweets and oily food which cannot be overcome by any system of discipline, so that the individual will not indulge in them when opportunity presents.

Bees fed on pure sugar refuse, for a few days, to forage among flowers. That, however, is only temporary, as we have often repeated the experiment, and found that, after the syrup had been mixed in the honey-pouch of the bees, with the secretions of that organ, it lost its fluidity by evaporation, and left dry sugar in the cell.   The bees immediately went earnestly to work, with united force, on discovering the appearance of things, and carried it all out of the hive, grain by grain, and then resumed their accustomed avocation in the fields.

When pigs are fed exclusively on boiled potatoes, though rich in starch, they fatten slowly, because no oily material is present.   By simply mixing milk with potatoes or corn-meal, seeds or nuts, the fattening process is vastly more rapid.

### Cooking Food for Animals.

Nothing is gained for domesticated animals by cooking their food.   Cows, fed on warm swill, still-house waste, macerated hay; or swine, urged on to excessive fatness by confinement and cooked food, have ulcerations of the liver, and a bad state of the tissues.   Meazly pork is a disease of the cellular texture, and, therefore, wholly unsuitable for the table, however disguised by pepper in sausage-meat, or bacon.   When Majendie rationed dogs wholly on starch or sugar, they died unexpectedly soon.   Butter or lard, fed to them exclusively, was equally fatal.

A duck fed entirely on butter, at the rate of 1,350 to 1,500 grains daily, died in three weeks.   On examination, the butter was oozing from all parts of the bird's body.   Even the feathers were saturated with it, and the odor was excessively nauseous and offensive.

### Warnings against Vegetarian Reforms.

The lessons taught in these experiments are detached evidences of a fundamental law of life, which cannot be set aside.

Persevering attempts of vegetarian reformers to convert men and women to their theories, propped up by the representations of the saving to be realized by returning to acorns, never have succeeded. Every little while a new aspirant for fame springs into transitory notice, to melt away under the sunshine of reason. There is no lack of converts when a new dietetic doctrine is first announced. There is a kind of romance in subsisting on next to nothing. Instead of needlessly wasting precious hours of a short existence in roasting legs of mutton, making pancakes and pudding, the whole twenty-four devoted to rejoicings over a glorious emancipation from the restraints and refinements of a burdensome civilization, is more poetical than profitable. With all the enthusiasm which usually characterizes the ardor of new disciples to any ultra proposition, the vegetarians fall from grace, and ultimately sin against arguments that were plausible enough at first, by returning to their former habits of living like sensible beings, in conformity to the usages of society in which their destiny is cast.

How ridiculous to attempt reasoning men and women into a conviction that their five special senses are not to be gratified, because it is displeasing to their Creator to indulge in anything he has bountifully supplied, simply as temptations, but not for consumption.

Of all modern reformers, vegetarians have the most discouraging prospect of success. It is recorded that Paracelsus prided himself in having discovered the true elixir of life. It was an expensive preparation which only kings could purchase, in expectation of living and ruling for ever on earth. While glorying in his pride, that his researches in occult science had terminated so favorably, he died with a bottle of his life-preserver in his pocket, at the age of forty!

## On What should we Subsist?

Not waiting for an echo to answer the question, reason says, whatever relishes. Any arbitrary system that prescribes positive rules and articles, to the exclusion of all others, must be wrong.

Dyspepsia was never cured by a spare diet. The false but fashionable direction for these whining, complaining, gaunt appendages of society, who are dying of indigestion, is the sure way of hastening their departure to that bourne from whence no traveller returns.

Dyspeptic invalids, besides slowly starving themselves in the midst of inviting plenty, pretty uniformly are all the while under medical treatment which is not required.

Assisting nature instead of thwarting her behests, by faring sumptuously every day on the bounties a kind Providence provides, offers a far better prospect of relief and a speedier restoration, than the slow, miserably wasting-away course usually pursued by intelligent sufferers.

This method of treating dyspepsia, the disease of comfortable circumstances, is no violation of the rational laws of health. Meet the malady with appropriate nutrition. Food for dyspeptics must be neither too fine nor concentrated, but a generous variety and of the best quality.

Those poor men and women who rarely gratify their palates with rich preparations which greet the uncertain appetites of the rich, are exempt from their peculiar sufferings.

Laboring people rarely ever have a symptom of that bane ,of pecuniary independence,—dyspepsia. They sleep soundly, and awake refreshed.

## FRUITS.

Fruits should be more freely used. Apples, especially, are exceedingly grateful to most persons. They may be cooked in many ways for the table, contributing largely to good living. Simply baked, they are excellent food, and, if eaten freely with every meal, act very beneficially on the stomach.

Fruit-eaters have health. Apples, pears, plums, peaches, berries, and, lastly, melons, may be eaten with impunity, if fully ripe. Children should not be denied, but allowed to revel in all the fruits in their season. They meet certain conditions, and, if withheld, the danger is far greater by a denial than from surfeit.

Parents are quite apt to limit children in the amount as well as the kind of fruit, on a presumptive theory of their own, that this or that would be injurious.

That is altogether a mistake. Crude fruits do derange the bowels, and produce disastrous consequences. But from ripe, unless they engorge themselves beyond the capacity of the stomach, no harm need be apprehended. Give children all the fruit they want. If it were not proper for them, they would not manifest an insatiable relish for it.

## CIDER.

When cider was a table beverage all over the apple regions of the Eastern States, forty or fifty years ago, there was a higher standard of family health than in these temperance times. When the temperance reformation was inaugurated, cider was anathematized as vulgar ; besides, its tendency was to stimulate, and, therefore, it must be dropped. It disappeared, and reappeared in the form of apple brandy. After the denouncement of homely, honest cider, which facilitated digestion, and kept up the strength of those hardy men who laid the foundation for the agricultural beauty and wealth of New England, dyspepsia made its appearance.

In those good old times, when honest men dealt honorably, cultivated their farms, paid their taxes, and brought up their children to respect all laws, divine and human, physicians were rarely called to their families. When the cider went, dyspepsia came in at the opposite door.

Malic acid facilitates digestion, without leaving any of those bad effects which follow the use of distilled liquors. Cider refreshes without leaving a sensation of lassitude, or disturbing the nervous system,—taken, as was formerly the custom, with dinner and in the field.

There were, occasionally, hardened old cider-drinkers, who took it immoderately, so as to be remarked upon as simply ridiculous, but drunkards are a later race. They came into notoriety with the multiplication of distilleries and the unpopularity of cider.

Physicians are guilty of a great moral wrong, by encouraging the use of whiskey, the curse of this magnificent country, where man alone is vile. Were dyspeptics to adopt cider as a diluent of their food, and totally abstain from tea, coffee, and, above all, whiskey, brandy, and wines, they could not be worse for it, and might regain their health.

There must be caution in the purchase of what is sold for cider. It is now manufactured extensively out of anything but apples. It is sold under the name of champagne cider, and that, too, is an outrageous imposition, and a dangerous compound for invalids.

The true medicinal cider—that which a dyspeptic lady or gentleman might take by the tumbler-full several times a day—should be such as is put up in barrels by the farmer in the interior, who is ignorant of the cheating ways of trade. Drawn from the barrel as it is to be used, and never permitted to stand till it becomes stale and loses its effervescent smartness, it will accomplish all that is claimed for it in this plea.

In these generalizations, in reference to a very common condition of ladies of middle age, and sometimes in young ladies whose lives have been too artificial, we have urged a new way of meeting their thin, shadowy forms, pale faces, attenuated arms, flat chests, hollow cheeks, and lassitude. Exercise on foot, indulge in luscious fruits, take less tea and concentrated food, and, by all means, patronize good, fresh, effervescing cider. The farmer's daughter escapes dyspepsia till she resides in a city where physicians are as plenty as lamp-posts, but not always as useful in showing the way.

## EXERCISES.

Proper exercise in the open air, which has been urgently recommended in these pages; an elastic, light bed, in a properly ventilated dormitory; early rising, if the lady has no further inclination for sleep; occupation alternating with agreeable amusements in the society of friends, or books; and always keeping physicians and drugs so distant as to be seen only through a telescope,—would bring feeble women, and pale, slender, drooping girls into the fold of Hygeia. Women have great need for making an effort, for they not only have very much degenerated, but they are further deteriorating, especially in cities.

Resolve to rise above indolence; and instead of reclining in an easy chair, with an India shawl over the shoulders, occasionally tasting with a teaspoon some delicacy, and when the clock strikes, very punctually taking either drops or pills—discard the whole of them.

Arouse from the insidious lethargy that holds you in its
folds, and face the breezes on foot many times in twenty-four
hours. When fatigued by long walks, take a refreshing nap;
next some substantial refreshment, and at reasonable intervals, repeat the exercise; when the weather is unfavorable, over-
see the house, look into the larder, calculate what will relish
for next day's dinner.

## TIMING FOOD TO THE SEASON.

Fruits come to maturity at precisely the period when they
are most serviceable. In their perfection, when their juices are
fresh, and grateful to the palate, the system is immensely benefited by a free use of them. It is not material whether a peach,
a melon, or a cluster of grapes, is taken at break of day, with
breakfast, at noon, night or midnight. When the stomach
craves them, it is the time to feast upon them. Still it is heralded from sources respected as oracular, by those who never
think for themselves, the eating of fruit should almost be regulated by statute law.

Those persons for whom no one seems to care, those who
get what they can, and when they can, unrestrained by arbitrary
rules in respect to living, suffer none of the predicted evils from
satisfying their appetites at any hour.

It is simply convenient to have specified hours for meals,
because an orderly system is introduced into the arrangements
of a family. There is economy of time in having regular hours
for all employments. An established habit of dining or sup-
ping at any particular hour, educates the stomach for that
period. Any marked deviations from a habit disturb its functions, simply because the digestive organs are not ready, or having been so, and not being provided for, ruffles the temper, quick-

ens the pulse, and thereby produces nervous irritability. Fruits are of such inestimable value in the maintenance of individual as well as public health, efforts should be made, particularly in compact cities, to provide the poor with it on a scale of liberality never yet inaugurated.

Poor children seize upon unripe and decayed remnants with a ravenous desire for them, as the season approaches for their appearance in market, which quickens the death record enormously; but ripe fruits correct and fortify the system just when a summer atmosphere is charged with elements that require counteracting agencies abounding in ripe fruits.

Benevolent schemes for ameliorating the circumstances of the poor will not be complete till some kind-hearted Crœsus provides for supplying them with generous supplies from the advent of strawberries to the gathering of grapes in autumn.

A benevolent Frenchman, Monsieur P. P. F. Degrand, left a handsome sum at his death, in Boston, the interest of which is to be annually expended in picture books for poor children. Besides gratifying the curiosity of the poor little recipients, who, otherwise, could never possess such a treasure as one of those instructive works appears in their estimation, they develop a love for reading, cultivate their taste, and bring out the first desire for improvement.

Fruits are always dear in this very fruitful country. The production has never been equal to the demand. Perhaps cultivators never wish to have them, as it would interfere with their profits. It is certain, that an acre of ground devoted to the growing of almost any kind of fruit would yield a far larger revenue than corn, potatoes, or grain, requiring a severe expenditure of labor in ploughing, hoeing, and harvesting. Why are not hundreds of acres set with fruit trees where there is now not one?

The poor long for fruits they cannot have, on account of the

price, disproportioned to their means. They barely procure
what are called necessaries—which means beef, pork, etc.; but
it may be positively affirmed that fruit is quite as necessary, and
far more important to them in their season.

A relish for fruits is not an acquired one, but born with us
—and for the purpose of introducing acids, saccharine juices,
and delicious flavors into the system.

Farmers! raise more fruit, and let the rich distribute it generously in tenement-house, cellars, shanties,—indeed, everywhere,
in lanes and filthy streets, where the poor are doomed to reside.
It would arrest diseases, it would relieve sufferings, meet the
urgent demands of the sick and feeble, and stimulate them to
efforts for improving their circumstances.

Many imperfections in our civilization might be corrected,
politically and morally. We are a confederacy of meat-eaters,
without much regard to its quality or quantity. We all consume too much meat. Once a day is enough in this climate.

Fruit-raisers are vehement in their assertions that it is not
only an unremunerative branch of industry, but there is also a
danger of over-stocking the market. There is not the slightest
prospect of overdoing the business. Since the process of preserving fruit is thoroughly understood, not a peach need be
lost, or a pear allowed to decay. The whole world over, they
are regarded as luxuries, and have a sure sale. If there is anything to be apprehended unfavorable to the fruit-growers' interest, it is that his avarice may urge him to ask more than they
are worth. Fruit-extortioners require rebuking.

## To be Encouraged.

As eminently contributing to the stability of public health,
and to the every-day comfort and improvement of the people,
the use of fish and fruit should be encouraged and upheld by

special laws.  In China, the consumption of fish is amazing; and nowhere is the public health, considering the denseness of the population, more satisfactory.  If rice and fish, the staples of life there, are reasons why neither plagues nor endemics are common, they might enter more freely into our own dietary with manifest advantage.  The Chinese are strong, well developed, and possess extraordinary powers of endurance. True, they require prodigious quanties of rice twice a day, fish being scarcely more than a savory relish, although they consider it essential to strength and vigor.

An educated Chinese brain, even under their objectionable civilization, is abundantly able to cope with the best diplomatic skill of Europe.  Much as they are underrated, their government antedates the oldest in Europe by thousands of years; and many of the useful arts and important discoveries— pillars on which the proud edifice of modern institutions are sustained—of incalculable importance to progressive humanity, actually originated among those Mongolians, whom we are taught to believe our intellectual inferiors.  We may not drink as much tea, stow away as much rice at a meal, or be as well satisfied with fish at every meal, yet they are by no means to be undervalued for their attainments in art or government.  Both are venerable for age.

### TOO MUCH MEAT.

Women with us consume too much meat—the result of a mistake in the beginning.  Neither the severity of the climate, nor the necessities of their systems, require it in large quantities.

Their indoor employments, with few exceptions, are such, a lighter and more easily digested food than meats would be better for them.  Farinaceous articles, including an abundance of fruit, fresh, cooked, or preserved, should be provided in all well-regulated families, especially where there are female children. Eggs and fish are proper, and avoiding pork always. Mutton is the most wholesome next to good beef. Sparkling eyes, an elastic step, elegant figures, a good temper, and quiet deportment, depend essentially on the food we are habitually consuming.

Irritability,—a desponding, dissatisfied state of mind, which gives a false coloring to nature, and makes women dissatisfied with themselves, and with all with whom they associate,—may be often traced to their improper food.

It is their mission to keep man, who is prone to displays of passion and outbursts of rage, in a bearable condition, by their talismanic presence. They would not be bearable even to one another, were it not for the magnetic influence of woman, who is the agent of all civilization, and certainly of refinement and morality. Even when silent, she rules the storms of human fury, and calms the savage exhibitions of wrath in men, by the charms of her character.

To succeed, she must neither dine on pork, nor inflame her blood with heavy, indigestible aliments.

Finally, less animal food than is now customary; abstinence from all heating, fiery drinks, which are never necessary for man or woman; varying the diet, so as not to become weary of any particular article or composition,—would improve us. It would give young growing girls a robust constitution, provided there is no limitation to out-door freedom. With such simple means, the women of this country may be regenerated; and their successors, the mothers of the coming men of renown, would be sound in body and strong in mind.

# CHAPTER XIV.

## INTERNAL STRUCTURE OF WOMEN.

Character of the Chest—Compression of Blood-vessels—Healthy Children—
Their Management—Scheme of the Circulation—Effects of Anger—
The Heart—Its Irritability—Origin of its Power—Sudden Death—Be
Moderate—Dropsical Effusions.

THERE is no apparent difference in the form or functions of
the viscera of the chest, or in the structure of the stomach and
its appendages, in men and women.

In consequence of the cramped position of the inferior ribs,
forced mechanically out of the line of natural incurvation by
stays, it is possible that the shape of the lower portion of the
lungs might give a clue to the sex to which they belonged, in
a judicial inquiry where that point was a question.

The chests of young ladies in our time, and in all Christian
countries in which there is an upper class, are trained with
quite as much care as gardeners bestow upon running vines to
give them direction. An experimental effort, to determine
from whence a pair of lungs were taken, might be decided by
the distortion of the bones about the cavity from which they
were detached. On the supposition that no interference with
the bones had ever occurred, neither exterior nor interior struc-
tural appearances would be any guide in reference to the sex of
the individual.

It is barely within the limits of possibility that a great crime
might require a decision in answer to a judicial question, Were
these the lungs of a man or a woman? A key for unlocking a

mystery is to be found by a simple examination of the inferior
margins of the lobes.

In a normally developed chest there is breadth at the base;
whereas, in artificially shaped ones, the lower part, which should
be roomy, is contracted, which obliges the lungs to conform to
the cavity in which they are lodged.

The lungs must necessarily expand with each inhalation of
air.  If the pulmonary cells are unnaturally small in one section
of the lobes, others beyond the sphere of restraint, by reason of
outside bands, will enlarge to abnormal dimensions higher up.
Surface is essential for the aeration of the blood.   If that process
is imperfectly accomplished, vitality is either quickly reduced,
or may never have been fully developed after the body was
fashionably put into harness.

A pale skin, feebleness, unsound health, are the penalties
for tampering with such delicately organized tissues as enter
into the composition of the lungs.

Between the extremities of the superior ribs—seven in
number on each side—the breast-bone, in children, is made up
of several distinct pieces.  Through all the early periods of
childhood, it may be readily forced from its normal relations
by keeping up a continual pressure in front.  The sternum, or
breast-bone, is simply a front wall, while the ribs and spine are
lateral and posterior protections of the contents of the pleural
cavities.

### MODIFYING THE CHEST.

Being never firmly ossified, even in advanced age, in females,
it is always in danger of being injured by their modes of dressing.  Women can be remodeled, on coming from the studios of
nature, under the plastic hand of the goddess of fashion, to
almost any pattern.

By lacing the chest in unelastic corsets the form is materially changed, always to the injury of the individual. But that seems not of the slightest consequence, since to live, breathe, and have a vulgar form, which the Divine Artist gave to humanity, has been entirely ignored by our refined, chaste conceptions of what female humanity should be, to meet the approval of cultivated taste.

The lower pendant extremity of the breast-bone (*zyphoid cartilage*) is quite flexible. If garments are tightly fitted to a waist already warped inwardly, to diminish its transverse diameter, the cartilaginous point is forced further inwardly, so as to encroach on organs lying directly behind.

Some years ago, in the course of daily lectures in a school of medicine, it was discovered, incidentally, that the skin was abraded and extremely red over the pit of the stomach of the female subject upon the table.

Evidently, there had been severe blisterings, which indicated some local difficulty that external irritants were intended to relieve. A history of the case could not be obtained. An exploration revealed the fact that the lower end of the breast-bone had been so pressed upon by force from without, as to bend it almost at a right angle. It was actually pricking, as it were, perpetually. Internal inflammation resulted, and no doubt the patient had suffered long and intensely from a deep-seated pain which no treatment could relieve,—it being, literally, a thorn in the flesh.

Both the pancreas and considerable of a patch of the under surface of the diaphragm had become diseased by being near the engorged vessels.

This illustrates the danger that may ensue by interfering with a living body regularly and harmoniously performing its functions.

It is quite familiar to surgeons that when an artery is enlarged into an aneurism, if one side of it touches a bone, gradually the solid structure will be removed by absorption at the point of contact.

Bones will not resist continued pressure without exhibiting disturbance. Therefore all appliances unfortunately imagined to improve the female form, even when quite gently commenced on the chest, are positively reprehensible. Girding the chest when the bones are imperfectly ossified, is extremely dangerous.

Swathing the frail, imperfectly made bones of newly-born children with bandages, rollers, or bands, a custom of almost universal practice even among intelligent mothers, on the mistaken idea that their backs require some support,—is worse than barbarism.

It is as absurd to swathe a new-born babe as the Indian custom of lashing them to a piece of bark, to make them straight. Civilized cruelty inflicted on an unresisting infant is a crime, which, in a more advanced state of civilization, may become an offence recognizable by the law.

Besides irritation of the skin, many a suffering child has been sent screaming with torments into eternity through the well-meaning intentions of an affectionate mother, who would have felt herself guilty of the sin of neglect had she failed to begin to make her child beautiful while its body was flexible and yielding.

Elastic flannel bandages, especially made to be easy, are abominable inventions. Cotton swathes, or any other bandaging, is a dreadful source of annoyance and misery to a nursing babe, of which they would loudly complain in tones far louder than crying, if they could speak of their misery.

Swathed from their arm-pits to their hips compresses the

blood-vessels; prevents the action of muscles that ought to be continually exercised, and must in the nature of things be a torment an adult would not submit to, even in stays, were it not for the impression that those who are thus self-tormented, are making their forms more agreeable objects for other eyes to contemplate.

## MUSCULAR FREEDOM

Perfect freedom of body should be granted the child from birth. No restraints, not absolutely necessary for cleanliness, should be imposed. Poor little things, they are dosed with nauseous drugs, made to swallow composing-drops unwillingly, and killed by well-intended measures for improving their forms. All the anxieties and difficulties attending the rearing of children, might be avoided by simply letting them alone. The poor raise large families successfully, because they have no time to spare in killing them by attempts to undo what nature will do, if not meddled with while engaged in perfecting her beautiful designs.

Children come into the world with all the machinery of organic life new and perfect. The mother's milk, which is their due, and not that of a hired nurse, contains precisely the materials for increasing the dimensions of the whole system and providing nourishment for each individual organ. Civilization, however, is not satisfied with appearances, and immediately commences schemes for improvement.

Like some unskilled artisans who, overrating their own acquirements, often spoil what they vainly attempt to improve, —so children that would have lived are victims to rude attempts to better what the Creator pronounced good when it left the laboratory, where it was fashioned in marvellous beauty and perfection.

Infantile bandaging, commenced when the bones are ductile, is the beginning, oftentimes, of a narrow chest, which would have had ample dimensions, had it not been tampered with before the framework of the skeleton expanded into full proportions. This civilized cruelty is the origin of an enfeebled constitution. If no interference were practised with a determination to alter the shape which would have been developed, the physical condition of woman would not be so generally defective as it is now known to be.

Were children from the first permitted to breathe uncontaminated air, by being removed from the too frequently vitiated atmosphere of an over-warmed nursery, nurtured on the mother's milk, instead of that of another woman's, whose physical and moral condition are entirely different, the child would present, in all its after-life, a very different condition. Milk from another source, although secreted in the breast of a healthy nurse, may introduce into the structure of the babe elements that immensely modify its original constitutional circumstances.

### A MOTHER SHOULD NURSE HER OWN CHILDREN.

Here is the gist of the whole matter. If we are to have beautiful and healthy children, the mother must nurse her own babes. Very many mothers who have no milk for days, or even weeks after confinement, under the impression that no secretion will take place, abandon attempts to promote it, too hastily. By repeated solicitations, allowing the infant to draw, as though the lactic flow were intact, stimulates the gland, so that milk rarely fails to come by patient perseverance.

Fresh cow's milk, especially that from a young animal having a calf, is safer to feed the infant upon till it appears in the fountain prepared for its secretion, than to furnish it from a wet-nurse, whose age, temperament, mental, physical, and even muscular condition are totally unlike those of the mother.

Leave off all swathes and bandages: that is the second important lesson to be remembered. Trotting young children violently, when they cry, to quiet them, is a fearfully reprehensible practice. Their frail bodies cannot bear such violence without endangering internal organs, by actually tearing away their attachments, and producing inflammations. Indeed, it is always hazardous to throw them about in the lap, as customary with nurses, without the slightest reference to their tender age and unfinished anatomy.

By allowing infants to lie on soft beds most of the time, till their spines are sufficiently strong to support them in a sitting posture with their playthings, in very loose clothing, unsmothered, in airy rooms, always sleeping alone, the next generation of women in the United States would be such beings as Nature intended,—fair, sound, and intellectual.

# CHAPTER XV.

## OVER-WORKING THE HEART.

### Value of Rest—Heart's Irritabilty—Arteries—Circulation—Influence.

SUDDEN emotions derange the functions of the heart. No persons are more familiarly conversant with the effects of painful or pleasurable emotions, or the extraordinary influence of sad or joyful intelligence, than woman.

Every one's experience furnishes conclusive evidence of the reflex influence of good or bad news, and the varying pulsations of the heart, resulting from mental impressions. When two beats are made in consequence of some sudden emotion, the contractility of the organ being quickened to perform twice the service it usually does in the same measure of time, it obviously tend to its injury.

In lesions, engorgements, abnormal depositions of fat within the pericardium, or the valves becoming slightly ossified, so that the auricles and ventricles are imperfectly closed, the administration of medicine is nearly useless.

How is it possible that a drug in the stomach, however potent in character, can remove a mechanical obstruction within the cavity of the heart?

Rather than retire from the turmoils of business, or fashionable excitements, or striving for social or political positions, diseases of the heart are multiplying. They are not produced by ordinary circumstances, but are generally the result of excessive effort in some direction for the attainment of an object, worthy or unworthy, which accelerated the activity of the heart,—a forcing engine on which life depends,—beyond its capacity. No permanent relief need be expected in the shop of an apothecary. There is no balm in Gilead for an enlarged heart, made so by compelling it to labor too much, or too long, at a rate beyond the motion it has when no unnatural stimulus has hastened its systole and diastole,—a succession of relaxations and contractions, which are natural and safe. Unnatural movements endanger its mechanism, especially if often repeated.

## VALUE OF REST.

Rest is a far better remedy for any irregularity in the circulation than medicine. Removal from the scene of excitement, and being out of the way, and beyond the sphere of associations or things which recall emotions that quicken the heart's action, is the true way of giving relief when diseased.

Where there are no extraordinary occurrences, but each day is a calm reproduction of the past—where broad fields, grazing herds, twittering songsters in the trees, and outgushing flowers invite admiration, and the contemplation of nature in the quietude of rural life, there should patients with irregularities of the heart take up their residence.

It requires a nice power of discrimination to determine whether a palpitation is caused by some interior difficulty, as for example, a thickening of the margins of the valves, ossification, obstruction in the coronary vessels, or arises from nervous debility.

In the latter case, the muscular power runs on uncontrolled, when the nervous power is weak, or nearly exhausted. Thus, after great fatigue, cramp seizes the limbs, the muscles contract spasmodically and irregularly, simply in consequence of

nervous exhaustion.  Sleep, food, friction, and stimulants replenish the battery, and then the muscular force of the voluntary cordage is perfectly subservient to volition again.

## THE HEART'S IRRITABILITY.

With an endowment of a kind of vitality peculiar, and, to
some extent, independent of all connection with the body, beating and throbbing when completely detached from the chest,
the heart is a wonder in itself.  It is the first to live and the
last to die.

Laid upon a table, unconnected by either nerves or vessels,
the heart of a reptile will expand and contract by the touch of a
pin.  Though blood is its appropriate stimulus, it dies gradually,
but may be partially revived by the introduction of air, or the
point of a needle.

The vital tenacity of the human heart is equally surprising.
It will withstand violent assault, deep wounds in its substance,
and encroachments of disease, far longer than would be supposed, were it not for revelations of morbid anatomy, which
occasionally demonstrate under what strange mechanical derangements it can sustain life.  Still it is a mortal machine, on
the regularity of which depend life and health.

When the heart fails prematurely under the pressure of
undue excitement, death is an inevitable consequence, which
neither skill nor science can avert.

To determine the amount of derangement in the system,
if any exists, physicians feel the pulse at the wrist, by pressing
the radial artery against the bone.  The number of beats there
corresponds uniformly with those of the heart.  Being telegraphed through the fingers of the examiner, intelligence
reaches the brain, where they are diligently compared with his
watch.

One large vessel carries all the blood forced from the heart. By giving off branches, which ramify extensively and minutely, the most distant fibre receives a proper amount of the vital fluid.

## ARTERIES.

Those intricately ramifying tubes, finer than hairs in their ultimate distribution, furnish blood from the centre to the whole circumference, in which, held in solution, are properties for the growth and reparation of whatever it passes through, over, or among.

When those soluble vitalizing elements have all been left along the track, according to the needs of each and every part, the blood then passes into the extremities of veins, by which it is collected to be returned to the right side of the heart.

The blood goes out of the left venticle, from the left side of the heart, of a rich red color, but it comes back to the other side of the heart, of a dark bluish color.

When the ventricle is fully distended, the walls of the heart suddenly contract with a twisting motion of its fibres, forcing the bluish blood through the pulmonary artery into the lungs, where it is brought in contact with atmospheric air, from which, in the twinkling of an eye, it absorbs its oxygen, gives off carbonic acid gas, and then plunges into the left side of the heart to repeat its rounds again.

## CIRCULATION.

Thus the blood is going and coming unceasingly from the first pulsation the heart ever made in its elementary, unfinished condition *in utero*, till its last beat, a death-kell at the close of life.

When the heart pulsates too slowly, or too rapidly, the physician forms an opinion, decides upon the character of the disease for which his advice is sought. He ought to be so thoroughly instructed, the least deviation from a normal standard of health may be quickly recognized.

In this climate, ordinarily, the heart beats from about sixty-five to seventy-eight strokes in a minute. Some, with the aspect of sound health, have only sixty, or even fewer, and there are others in equally sound condition, whose pulse habitually exceeds eighty.

A pulse, however, varying through the whole twenty-four hours, according to the stimulant effects of food and drinks, does not indicate sickness. There may be a sudden alarm, through the acoustic nerve, the instantaneous apprehension of danger through the optic nerves, by the sight of the edge of a precipice, a falling rock, an approaching wave, or terrific explosions of thunder, or the flashings of lightning in the sky, which may instantaneously increase the action of the heart to more than a hundred strokes. Through the nerves of sense, so great has been the shock that the heart has burst.

### INFLUENCE OF ANGER.

Extreme paroxysms of anger are sometimes fatal by an explosion of life, as it were. The heart resists spasmodic demand made upon it to empty its cavities, and bursts. Rents in its walls, which are almost instantaneous death, have often been found produced by extreme exhibitions of rage. It is always dangerous to indulge in unrestrained wrath, especially for women of a nervous sanguine temperament.

With some, the pulse is preternaturally rapid. Others are equally remarkable for the moderation of the heart, always moving at a very nearly uniform rate.

Blood, which is a vital fluid, is driven through the arterial canals at an average velocity in health. It is neither hurried nor retarded by trivial circumstances. When the heart beats a hundred times in a minute, it is a sign something is wrong, if it continues for a considerable time to throb and labor thus actively. When by treatment that rapid action cannot be moderated, death's messenger is in waiting. With all the poetry with which the human heart is invested, it is simply a forcing-pump of immense energy. Instead of being kept in motion by exterior stimuli, it contains within itself contractile fibres, which are obedient to the contact of blood. Its presence in the interior of the organ calls into action a mass of winding muscular threads, whose combined contractile force is equal to the grip of a strong vice, in expelling the current that has just arrived.

A relaxation succeeds the violent contraction of the walls. For an instant, those ever-working muscular filaments rest, then resume labor again.

### The Heart a Double Forcing-Engine.

More critically considered, we really possess two hearts. One of them belongs to the lungs, while the other is for the body. They are joined together, and, therefore, have the appearance of a single organ. Nature invariably economizes room. By uniting the two hearts, the necessity of having separate apartments was obviated, when one would answer all purposes.

In some reptiles, the two hearts have been found separated. We have an indistinct recollection of having read of a case in which the two hearts were at considerable distance from each other, in a patient carried to an European hospital.

One heart receives all the deteriorated blood, by which is understood that gathered up and brought to the right heart, having left its life-sustaining properties in passing through the body. Being again forced into the lungs by an immensely powerful forcing-pump, it there again imbibes oxygen from air waiting for it in the cellular structure of those membranous sacs. From thence it is again forced into the upper part of the left heart, on the left side of the chest, and next into its ventricle, more powerful as a forcing-engine than any of the others, which drives the living current into a single elastic tube, the aorta, to pursue its mission through the system again.

The irritability of the heart, from the earliest embryotic condition to one hundred years—and in Henry Jenkins, one hundred and sixty years—is not well understood.

Two French physiologists have announced the discovery, says report, of two nerves that have heretofore escaped the inquisitive researches of anatomists, creeping out from the side of the vertebral column, which ramify extensively in the tissues of the heart, and through their instrumentality the motor power is kept up.

A certain Dr. Cyon, of France, has sent forth a learned dissertation on the heart's innervation, explanatory of the function of those newly discovered cords. One of them is recognized as the accelerator, and the other the motor nerve.

How it happens that a heart pulsates when severed from its connections entirely, for more than half an hour, makes the problem of its independent vitality more abstruse.

### EXCITABILITY.

As a people, we have a reputation for being always in haste. As a consequence of this hurrying propensity, both men and women wear themselves out prematurely. Merchants are over-anxious to be rich; ladies, too, ambitious beyond reason, over-work their hearts.

Sudden death from heart-disease is a common coroner's report. Juries of inquests have not assumed the responsibility they would be justified in taking, by a verdict of over-excite-ment of the brain, or over-taxing the heart.

Competition in trade, deferred hopes, unexpected disap-pointments, pecuniary losses, a reckless determination to carry measures which are extremely hazardous, often resulting in disastrous failures, shock the nervous system by a reflex action upon an over-excited brain that recoils upon the heart.

A familiar expression—*broken heart*—is not inappropriate. They do break. Mental emotions may be so intensified as to produce paralysis of the heart. A fatal spasm of its muscular walls is induced from a sudden painful impression or fright. Sudden deaths from such causes cannot be reasonably doubted.

A fearful penalty of a violation of a law of health, is when a person concentrates too much will-power suddenly. Revenge or hate, while under the influence of stimulants or excessive politi-cal excitement, may end in instantaneous death from a spasm of the heart. When a contraction is accomplished under such circumstances, it holds its grip, and death closes the scene. Sometimes there is a rent in the flesh of the heart, through which a gush of blood escapes into the heart-case,—*pericar-dium*, and that is a death-lesion for which there is no relief.

Moderation in legitimate pursuits should be encouraged. "Be not too ardent" is a caution to be remembered, especially by youthful, sprightly, passionate young ladies.

Formerly, the heart was supposed to be the abode of moral sentiments. It has the credit of being open to amatory impres-sions, as the focus of the affections and the fountain of love.

When that idea was taught as a truth, the bowels were exultingly referred to as the real seat of compassion! Both theories were found to be erroneous; but the mistake had been so long and extensively propagated in poetical fictions, in the language of all nations, the heart and bowels have been permitted to keep possession of those two attributes, and we continue to appeal to the deep feelings of the heart, and the yearnings of the bowels.

Women are not quite so much prone to the development of diseases of the heart as men, because they are generally less exposed to violent turmoils which wreck the constitution. They, happily, are removed from arenas of political strife, and from dissipations that make the blood boil. They never haunt drinking-saloons, those plague-spots of a city, nor carouse through the night in boisterous hilarity. They cannot, however, bear up under assaults upon their reputation, nor heroically defy slanders, without reeling under their crushing weight. Innate pride, the strong power of innocence and a consciousness of doing no wrong, sustains them awhile under such assaults, but they give way at last. They have dropped dead from a sense of injustice.

But women oftener rupture the heart by a paroxysm of dreadful rage, than from other causes. They have a safety-valve in a copious flood of tears. Under excitements that would explode life in some men, a woman is instantly relieved when the tears flow. They take off the tension.

When the brain is once charged with blood, by an increased action of the heart, by reason of exasperation, carried in faster than it is carried away by veins, an apoplexy would probably follow, were it not for immediate relief in a hearty cry.

Men breast a storm of passion better than women, but there

is no merit in it. They oppose whirlwinds with whirlwinds, and yield at last at the sight of a woman's tears.

Death from ossification of the valves or coronary arteries, those which immediately supply the heart for its own support, together with sudden paralysis, are far more frequent among men than women. Those maladies are on the increase. Merchants, bankers, speculators, and radical political leaders, who meet with damaging rebuffs just as they are expecting to win the prize, are those who fall suddenly dead.

Women have hearts preternaturally enlarged. They also are predisposed to have accumulations of fat around the organ, that impede its motions and mechanical regularity. Enlarged hearts may result from other causes, among which is excessive grief.

Disappointments, in which the affections are deeply involved, may be a cause of diminished vitality.

Dropsical effusions are apt to follow that state, accompanied by functional derangements.

An intermitting pulse, with an occasional twinge in the region of the heart, indicates, generally, in women, nervous debility, which may be aggravated by mental excitements or continued apprehensions of a calamity.

The reticence of women, their secretiveness, and the tenacity with which they conceal the causes of their unhappiness, when their pride is wounded or their preference slighted, obliges a physician to guess at causes very frequently. His prescriptions, under such circumstances, are random shots in the dark:

> " Earth hath no rage like love to hatred turned,
> Or hell a fury like a woman spurned."

# CHAPTER XVI.

## THEIR LUNGS.

WOMEN, oftener than men, do violence to their lungs. It may not be agreeable to be told they are habitually abusing those very essential organs.

It is a melancholy reflection that the progress of pulmonary consumption in this beautiful country is largely due to a vice in dress, which interferes with the development of the chest.

A residence in a crowded city, or, indeed, wherever there is a dense population, is attended with some degree of peril in respect to the purity of the air. If it is mixed, and charged with noxious vapors, or there is a deficiency of oxygen, animals breathing it cannot be in the good condition they would be in, in localities where no such vile elements were inhaled.

Consumption is alarmingly hereditary. Sporadic cases are also increasing, induced by causes which might be avoided to some satisfactory extent, if the demands of fashion were not so extremely arbitrary.

A sense of smell warns us of the bad quality of air in the vicinity of certain manufacturing establishments, such as gas-works; bone-boiling nuisances; slaughter-houses; putrefying carcases; decomposing vegetables, or other sources of impurity that would be injurious if inhaled.

Our olfactory nerves are special sentinels, promptly announcing sources of offence, and giving timely warning that they may be avoided.

## HEREDITARY CONSUMPTION.

Hereditary consumption is a hopeless form of that dreadful malady. Those influences, or agencies which bring on inflammation of the lungs, are comparatively few, compared with the annual devastation of human life from transmitted sources, propagated in families from one generation to another.

No sensible physician admits that pulmonary consumption is either infectious or contagious; while those knowing the least about the laws of disease firmly believe, as in Cuba, that it may actually be communicated by a touch of the furniture, or air of an apartment in which a patient with that disease has died. Hence, a theory sometimes assumes the dignity of a fact, and ignorance is better received as authority than scientific intelligence.

Medical authors assume it to be a firmly established opinion, that pulmonary consumption is a concomitant of modern civilization. While our ancestors, in the United States, occupied ruder dwellings, through which the air traversed freely, and they subsisted on plainer and coarser food, consumption was rare. With the advent of warm houses, coal furnaces, heated apartments, luxurious tables, and a tainted atmosphere, made so by imperfect ventilation, increase of population, domesticated animals, and manufactories of every imaginable description, the death rate has increased to an appalling degree.

Proofs are not wanting to show, also, that modes of dressing, imperfectly adapted to the varying temperature of the climate, is another prolific and very certain source of lung difficulties in females, which terminate in the ulceration and destruction of those organs. Indian habits at the West furnish abundant materials for determining many propositions respecting the development of thoracic diseases.

Those who are surrounded by domestic comforts, protected · from atmospheric humidities, or chilling blasts; who sleep in properly ventilated apartments, and are warmly clad at seasons when the weather demands special attention that perspiration shall neither be excessive, nor suddenly checked by exposure, are also subject to the same class of pectoral inflammations as those who repose on the ground in the smoke of a wigwam.

The diet of the Indian is mostly animal, and simple enough as far as it goes to meet the approval of an exacting stickler for plain food; and yet they die frequently of pulmonary consumption.

Dr. Rush assured his readers it was unknown to the aborigines of this country. He was eminent in his day; but more extended intercourse with tribes all through the interior of the continent since that distinguished author passed away, demonstrates the existence, and the melancholy ravages, too, of that plague among savages, quite as severe in proportion to their numbers, as where the resources of civilization are unlimited.

### PREVALENCE AMONG SAVAGES.

Red Jacket, the famous chief, whose name is interwoven in the web of modern American history as a wild man of extraordinary intelligence and political sagacity, assured a Buffalo physician about the year 1823, that no less than seventeen fatal cases of consumption had occurred in his own family, including ten of his children.

Other memoranda of a similar import might be given, conclusively establishing the fact that the disease has always been regarded by the Indians as incurable.

The reason why it is incurable, in its advanced stages, is because there has been a destruction of portions of organs, without which life cannot be sustained.

Aboriginal habits, customs, privations, and their brave darings in the chase, in war, and their ardor in feats of strength, must expose them to severe colds when heated or in a glow of perspiration. Lying down on the damp ground to sleep; wading through jungles, and shaded from the life-giving properties of sunlight by wide-spreading branches of trees in those forests where they prefer to roam, must lower their vital temperature and predispose them to the development of many painful and fatal maladies.

Sporadic pulmonary consumption, therefore, on reflection, seems to be most frequent with the Indians; while hereditary forms of it predominate in civilized society.

## VENTILATION.

Apartments may be satisfactorily ventilated by the latest patented contrivance, without essentially modifying the condition of the air in them, if it is laden with the products of low lands, noxious gases, or the putrid decomposition of animal remains. There is room for improvement in the management of wool and cotton mills, dye-houses. and gas works, so that they shall not interfere with the health of operatives.

Where large numbers of females are employed, further efforts should be made for giving them pure air for respiration.

In manufacturing establishments, especially in those where several hundred women are congregated, the messengers of death soon approach them in all imaginary forms, if ventilation is neglected. Females thus associated suffer more than men placed under similar circumstances.

Private residences, school-rooms, basement apartments, and stables are too much neglected in respect to fresh air. Where windows are not frequently opened and fresh currents allowed to displace those accumulations of dust, invisible spores of minute vegetations accumulate in an undisturbed atmosphere. Eggs of insects and impurities of various kinds destructive to health, generate also numerous diseases. In such conditions of air we oftentimes breathe, without being conscious of the existence of such subtle agencies. A lodgment of these microscopic irritants in the lungs are met by nature's only means of defence,—an extra secretion and pouring out of a fluid from a mucous surface to wash away offensive irritants.

### Tobacco an Offence to the Salivary Glands.

On that principle tobacco is an unwelcome injurious excitant, and the salivary glands pour out an immense amount of saliva to float off the obnoxious quid. When the effort is first commenced to chew or smoke, the quantity of saliva is more copious than after the individual has schooled his salivary apparatus to bear the presence of a terrible narcotic with some degree of acquiescence; but the glands never, at the end of fifty years, cease to manifest a dislike to tobacco in any form, by an increased activity of all the buccal and sublingual secretory organs at the instant it is introduced into the mouth.

Both smokers and chewers are constantly expectorating and spitting, to the disgust of those in their company, and certainly to the manifest injury of themselves.

## A Common Origin of Pulmonary Irritation.

In consequence of the lodgment of tiny particles of matter in the lungs, they produce a very slight irritation at first. A cough, however, is generally sure to follow, and that is simply a mechanical effort to throw off the irritant.

If the adhering atoms cannot be removed by a spasmodic blast of air from the lungs, then the next effort to overcome its offensive presence is by pouring out a large amount of adhesive mucus to entangle them, as it were, affording a better chance of expelling the intruders by acting on a larger mass. Thus there is a hacking expectoration.

Thus a settled cough may be produced. By constant repetitions, convulsive throes actually lacerate the air-cells, and ultimately involve the whole respiratory organs in disease.

When lesions become extensive, and one air-cell is ruptured, so that two, or three, or dozens become one cavity, the thick mucus collects in such quantity, besides being exceedingly tenacious, that a cough cannot raise it. The collection finally distends those delicate receptacles, more and more deranging contiguous cells,—and that is the formation of a pulmonary abscess.

By its weight and purulent character, respiration becomes not only painful, but hardly surface enough remains in the contiguous respiratory cells to oxygenate the blood sent to them to be vitalized.

This is the last and hopeless state of pulmonary consumption.*

---

* It is a well-recognized fact that the colder the climate, the higher the latitude, and the drier the atmosphere, the less liable the inhabitants are to suffer from consumption. In Iceland, from 1727 to 1837, there was not a single case, and Sir R. Parry, in his history of his northern explorations, noticed the rarity of throat and lung affections among the inhabitants of Greenland and Labrador.

It is not the object of this publication to provide a guide for the practice of medicine, nor attempt to persuade those who may honor it with a reading, that they can prescribe for themselves when sick.

---

In the two British stations of the Mediterranean, Gibraltar and Malta, long known as favorite resorts for the consumptive, we find the disease to be actually more prevalent than in Canada, with its long cold winter.

In Canada, six men per thousand of the British army are attacked by, and half that number die of consumption.

In Malta there are nine per thousand attacked, and four per thousand die of the disease. In Gibraltar the number attacked is seven, and the number of deaths three per thousand men.

In the Bermudas, where the climate is uniform, eight men per thousand become consumptive, and five of that number die. But in Newfoundland, the mortality from this disease is but four in ten hundred.

In tropical countries, the progress of consumption is more rapid than where the climate is temperate. Deaths from this ailment are more numerous in Brazil than in Russia. Owing to the extent of territory, and the different latitudes and climates embraced in the United States, there is, as might be supposed, a corresponding variation in the prevalence of consumption. We find the mortality from this malady to be greater in the New England States than in any other part of the Union.

The death rate by consumption in the States and Territories of the Union is shown in the following table:

| | | | |
|---|---|---|---|
| Alabama | 1 death in 25 | Mississippi | 1 death in 18 |
| Arkansas | 1 death in 22 | Missouri | 1 death in 26 |
| California | 1 death in 100 | New Hampshire | 1 death in 4 |
| Columbia District | 1 death in 6 | New Jersey | 1 death in 7 |
| Connecticut | 1 death in 5 | New Mexico | 1 death in 72 |
| Delaware | 1 death in 10 | New York | 1 death in 6 |
| Florida | 1 death in 21 | North Carolina | 1 death in 18 |
| Georgia | 1 death in 35 | Ohio | 1 death in 11 |
| Illinois | 1 death in 13 | Oregon | 1 death in 9 |
| Indiana | 1 death in 11 | Pennsylvania | 1 death in 8 |
| Iowa | 1 death in 11 | Rhode Island | 1 death in 4 |
| Kentucky | 1 death in 11 | South Carolina | 1 death in 30 |
| Louisiana | 1 death in 13 | Tennessee | 1 death in 13 |
| Maine | 1 death in 6 | Texas | 1 death in 27 |
| Maryland | 1 death in 8 | Utah | 1 death in 20 |
| Massachusetts | 1 death in 5 | Virginia | 1 death in 11 |
| Michigan | 1 death in 6 | Vermont | 1 death in 4 |
| Minnesota | 1 death in 29 | Wisconsin | 1 death in 10 |

The small proportion of mortality from consumption in California was

## NOT SAFE TO DOCTOR ONE'S SELF.

It is a maxim with lawyers, that he who pleads his own case has a fool for a client. Those who expect to be their own physicians, on the self-complacent notion that they understand their own constitution better than those who have been laboriously studying the morbid conditions to which humanity is incident, make a mistake which cannot be readily rectified.

To show how incipient forms of disease may be avoided, as well as caused, with plain suggestions respecting the maintenance of health, is of more importance to non-professional readers than a volume of recipes.

## MEDICAL IMPOSITIONS.

Beware of medical impostors. This country is an active theatre for the display of their peculiar talents. It is a profitable specialty to trade in advertised falsely-called remedies for consumption.

By baiting the trap, as a hunter would say, which is nothing less than encouraging a forlorn hope, those who have sought relief without finding it, purchase liberally and pay dearly for stuff that cannot accomplish cures when the substance of the lungs, or portions of them, are actually destroyed.

---

accounted for by the fact that the greater part of the population was composed of miners and emigrants from other parts, who were over 25 years of age, and not so liable to its attacks. More recent statistics have confirmed the assertion, that consumption is much more prevalent on the Atlantic coast than in California.

Daily variation in the temperature is believed to be the great cause of the excess of mortality in the Eastern States.

In proportion to the population, the number afflicted by this "destroyer of mankind," is frequently greater in small cities than in large ones.

Treat with contempt advertised certificates constructed for encouraging hopes that never can be realized.   Shun consumption doctors as you would seventh sons, clairvoyant seventh daughters, pickpockets, and professed swindlers.

Indian doctors!  those hypocrites and ignoramuses who announce themselves as having been taught by savages to do what men of science cannot do, is an absurdity.   No person of common intelligence believes one person can see further into a millstone than another.

If those who have studied the minute anatomy of the body, and have watched the operation of drugs in every possible phase in great hospitals, under the critical instruction of distinguished clinical professors, cannot arrest the destructive march of pulmonary consumption, is there any good reason for supposing that ignorant, vulgar pretenders, half of whom can neither read, speak, nor write their mother-tongue grammatically, possess knowledge superior to such as are educated under all the advantages of the age?

There are consumption curers entirely ignorant of the mechanical structure of the lungs, as they are of other viscera in the cavities of the body, who seem to magnetize those falling within the sphere of their operations, so that some very sensible people become their victims.

Consumption is an exhaustless theme.   Weak lungs or strong lungs are subjects for discussion when no such expressions are scientifically allowable.   Susceptibility to certain influences as sources of irritation to those delicate organs, is what is to be understood, and not that in the sense of a strong muscle, or a strong rope, or a strong beam, are they to be represented.

## CONTRACTED CHESTS.

Women, far more commonly than men, have contracted chests, which mechanically prevent a full inflation of the lungs to the extent they would be filled in a chest of larger capacity.

When air is simply inhaled, there is taken from it oxygen, —an element that sustains life. That being accomplished, the waiting air, thus deprived of one of its constituents, is forced out through the same tubular passage by which it was drawn in, carrying with it carbonic acid gas.

Such is the process and the object of breathing. By respiration, blood meets air in the lungs, where the exchange is made of something that cannot be safely retained, for that which maintains life.

Carbonic acid gas is taken up largely by growing vegetation, which they exchange for oxygen, that supports animal life.

With the cessation of respiration, the pulsations of the heart gradually terminate, and then unconsciousness follows. In drowning, those phenomena succeed each other in rapid succession.

## RESUSCITATION.

Left thus, an individual is popularly considered dead. But if quickly taken from the water, when all the functions of life are apparently forever ended—the heart no longer beating, the lungs collapsed, and consciousness gone—vitality may be recalled by persistent efforts.

Artificial inflation, the application of warmth, and the pursuance of directions extensively disseminated by humane societies, for the express purpose of informing people how to proceed for the recovery of drowned persons, often recall the apparently-dead to life again.

Such restorations are splendid triumphs of science. Alternately filling and pressing out the air from the lungs, by working the intercostal muscles, enlarges first the pleural cavity, then it is as suddenly diminished by the expulsion of the air, imitating natural respiration.

The air-cells are thus expanded to their full capacity. By continuing the process perseveringly awhile, the blood begins to absorb oxygen. As soon as that takes place, the heart feels the stimulus and contracts.

Through the agency of muscles thus manipulated, a reflex power is transmitted to both heart and lungs, and they then continue to act without assistance. The soul is recalled.

Where was the soul during suspended animation? Whence came it, by carrying on this mechanical effort, to bring the dead to life again?

### Value of Gymnastic Exercises.

Reasonable gymnastic exercises are exceedingly serviceable. The inner capacity of the chest may be very considerably enlarged by systematic exercise of the exterior pectoral muscles. The further an individual advances in age, the more difficult it is to overcome rigidity, or spread bones held closely by inelastic ligaments.

By commencing seasonably, before that condition is established, the conformation of the thorax or chest, which may be too narrow and too flat for a full development· of the lungs, may be very considerably expanded. Robustness and vigor may be attained, of the highest importance in regard to health and longevity, by simply compelling motor cords and strap-like tissues to pull back, out of the way of the swelling lungs, those too much incurvated ribs that prevent a full inhalation of air for filling the air-cells.

Ladders, inclined planes, swinging at arm's length in slings, climbing suspended ropes, pitching quoits, driving a ball, or following out the directions of acknowledged experts and public benefactors, who teach hygienic laws, to the saving of thousands of valuable lives that otherwise would long since have been entombed, had it not been for their valuable lessons, —is far more agreeable than emetics, blisters, tonic tinctures, or other products of a drug-store.

When lesions exist, there may be hemorrhages, or a tendency to expectoration of blood from a continued inflammation of the lining membrane of the bronchial tubes, indicating a condition that forbids gymnastic exercises.  It is then best for a person thus circumstanced, with graver symptoms to be apprehended, to change location.

## CHANGING LOCATION.

Avoid medicines, then, which are not decidely tonic, it being impossible to bear up under the action of drugs which have a sedative influence, or those which, like active cathartics, suddenly reduce the vital force.

In making a removal, it is essential to seek a residence where the atmosphere is dry.  Humidity is the bane of consumptives.

Sleeping over stables, with an expectation that evaporating filth from fermenting manure will heal ulcerated lungs, or strengthen feeble tissues in air-cells, is quite as unphilosophical as a residence in the Mammoth Cave for the same purpose.

St. Paul, Minnesota, has a reputation for being a hopeful temporary abode for consumptives, provided the patient is

prompt in going there before the disease has made that destructive progress which a change of climate cannot arrest.

It has been questioned by some medical men whether St.
Paul really does work the change which has been claimed for it,
as a resource for consumptives.  Possibly the journey from any
considerable distance contributes more directly to their benefit
than may have occurred to those who warmly recommend a
dry, elevated position.

Florida, also, has its advocates for the same class of invalids.
Many have been exceedingly benefited by a residence of a few
months there.  Avoiding the harsh, cold, damp winds and
easterly weather, of New England particularly, when the in-
clement winter of the Atlantic shores sets in, by escaping to
the mild regions of the South, must certainly afford relief to
diseased lungs, and give the general system some chance for
recuperation from that extreme debility which follows in the
train of a protracted cough.

Two miles from the mouth of that celebrated cave just referred to, remains of huts may still be seen, roofless of course,
where numbers of emaciated strangers in all stages of consumption resided in thick darkness, if their lamps happened to
go out.

Constant coughing and the repeating echoes of those sepul-
chral sounds that were forerunners of approaching dissolution,
together with smoke, which were as unendurable as their individual pains, soon destroyed the romance or hallucination,
whichever it may have been, and those who survived those
isolated trials in search of health in the gloomy bowels of the
earth, were glad to return to their inviting homes.

The theory which influenced consumptives to wend their
way to the great Kentucky cave, was that the saltpetred atmosphere in the interior was a remedy for ulcerated lungs.

Pulmonary consumption is everywhere. It is quite as well to remain at home, under certain forms of the malady, as to seek relief in other latitudes.

The little that may be temporarily gained by long and expensive journeys to some imagined place of restoration, is not a compensation for deprivations of society, and those friends and associations, devoted relatives and sympathizing acquaintances, medical attendants and familiar scenery, which are enhanced in value the farther we are removed from them.

## What to Do and What to Avoid.

Horseback exercise; all forms of gymnastic feats which give a wide range of play to the pectoral muscles, together with a generous diet, are always first to be tried in incipient forms of this particular disease.

Avoiding a free out-door exposure when the weather is clear and dry, is a mistake. Humidity, heavy dews, rain and cold, give activity to those processes of derangement in the lungs which hasten a fatal termination of life. Therefore it is important to sleep warmly protected, while there is a free circulation, or, at least, a free admission of air into the apartment, without fear of inhaling dangerous elements from that source.

Eating whatever relishes is not to be overlooked in a desire to take advantage of all available circumstances for promoting the comfort of a consumptive. There should be no restrictions in regard to food. The appetite is exceedingly capricious, therefore whatever is coveted may be taken with impunity. If oily food, butter, cream, fat meats, etc., agree with the individual, the more freely they are taken the better.

Systematically, that is, at regular periods, at suitable intervals, take cod-liver oil. Its value has not been overrated. For a time there was danger of its utility being undervalued in consequence of the general repugnance of patients to taking it on account of the disagreeable fishy smell, and the nausea induced by it in some irritable stomachs.

Happily for the reputation of modern pharmacy, cod-liver oil is now so admirably prepared, its objectionable taste is overcome, so that it may be taken without hesitancy,—all its unpleasant taste and odor being taken away without impairing its medicinal properties.

Cod-liver oil is not considered medicine, in the common acceptation of that term, but nutritious animal food that furnishes materials for repairing a wasted form.

Abstain from whiskey and similar heating stimulants. Physicians who have urged such treatment have done the country an irreparable wrong.

Unintentionally, they have made drunkards, by developing a morbid inclination for ardent spirits, which cannot always be overcome, when the discovery is made that the remedy is as bad as, if not worse than, the disease for which it was prescribed.

One of the simplest precautions for preventing inflammatory attacks of the lungs, is to be shod and clothed suitably. Ladies, particularly, invite death's doings, by being in extremely thin shoes, and light dresses that conduct off the caloric of the body, which should be retained by non-conducting clothing, when they find themselves threatened with a cough.

Thinly dressed, with the chest half exposed to direct blasts of cold air; standing at open windows in a current, or sitting out-door in a damp atmosphere, leaving a warm room for a cold one; dancing till heated by exercise, and then stepping into a carriage in a glow of perspiration, half protected by a silk cloak, a thousand dollar gossamer shawl, instead of a woollen blanket, are so many ways of inviting conditions of health which no medical skill is competent to manage.

The mucous passages, especially those leading to the lungs, are the first to suffer under such courses of imprudence. The lungs become engorged with blood when the lining membrane is flushed with a commencing inflammation, which rarely fails to be accompanied by a hectic cough.

### Violation of General Laws of Health.

Happily, women are beginning to discover the dangers that surround them, in conforming to the wild caprices of fashions. Those who escape pulmonary consumption by their violation of sanitary laws, are frequent sufferers from pleurisy, usually originating in the same kind of imprudence which generates other formidable evils.

### Pleurisy.

Instead of being confined to the lining membrane of the cells within the lobes of the lungs, pleurisy means an inflammation of the pleura, or living membrane of the chest in which the lungs play.

Whenever the inflammation becomes acutely painful in pleurisy, the attempted full inflation of the lungs must necessarily press against the inflamed surface. A stitch in the side, a common expression, simply means that the outside covering of the lungs has become attached or glued, as it were, to the membrane next the ribs—and the stitch is but tearing them apart—or rather, bridles of adhesive serous fluid, put upon the stretch, cause that acute sensation, a pain always attended with danger.

Instead of patronizing shoes, the soles of which are scarcely thicker than paper, it is quite as proper for females to wear them of sufficient thickness, as for men.

When the feet are cold and kept so for hours, in consequence of the waste of warmth through thin soles, the circulation of blood in minute vessels at such a distance from the heart, is partially interrupted. That cannot be habitually practised without deranging the general circulation. Swelled feet are the result of cold and compression.

The torture of tight shoes does not wholly consist in the development of corns and bunions, but in the production of conditions in the mechanism of the circulation that may degenerate into actual organic lesions.

Ladies should have their feet and ankles as completely protected as men who would soon be incapacitated for active pursuits were they put into the frail shoes and gossamer stockings, which are the pride of a well-dressed woman.

### SUSPENDED, NOT CURED.

Hereditary consumption cannot, with certainty, be averted. It may be suspended, as it were—or rather kept at bay by changing residence to a propitious climate. But all such measures are regarded as temporary. Nothing is more difficult than to stop the progress of a disease which destroys the organ by which life is positively sustained.

Sporadic, or that form of pulmonary consumption, induced by carelessness or unfortunate exposure to influences that could not, or would not, be avoided at the time, is to be managed differently.

By an imprudent exposure to cold and humidity, an impetus is given to the development of quiescent tubercles. They are suddenly inflamed, and suppurate.   In hereditary consumption, tubercles are actually found imbedded in the lung tissues of new-born infants.   They may remain many years perfectly indolent, if those precautions are taken which are pointed out in the foregoing observations, that have a tendency to awaken them from a long slumber into activity.

We do not believe hereditary consumption can be arrested permanently, so that it may not be transmitted to the children of such unfortunates.   But it is quite certain life may be considerably prolonged by a judicious reference to latitude and longitude, before grave symptoms indicate an ulceration of the air-cells.

# CHAPTER XVII.

## DIGESTION.

IT may be surprising intelligence to those who importune physicians as to what they should eat and drink, or what they might take into their stomachs with impunity, to assure them that medical practitioners are no better judges on that subject than themselves.

Because medical men are supposed to be laboriously interrogating Nature for information that may be of service to those who employ them, they are held accountable to a certain extent by a confiding public, in regard to the health of those who seek their advice.

Unfortunately, medical Solomons disagree among themselves. There is no standard by which to regulate the sanitary condition of society. They entertain theories enough to perplex all the universities on the globe; but the facts which always have precedence over speculations, are comparatively few, and not much relished by those who are ambitious for establishing theories as substitutes.

Digestion is a familiar topic, especially with persons profoundly ignorant of their own organization, and indigestion is still less understood by many who assume to be extremely wise. There is no definite system to be pursued, that will insure immunity from indigestion, by recourse to drugs.

Were we to say let medicine alone entirely, it might be

thought a selfish purpose was in view. Unhappily for those seeking reliable information respecting the course to be pursued to insure the highest standard of health, medical philosophers strangely disagree, so that invalids are perplexed, and, on the whole, derive about as much benefit from one source as another.

No one set of stereotyped directions meets every case of indigestion. There are no specifics for dyspepsia. Treatment that has been efficacious for one person, is of no service to another.

It is curious to examine the rules laid down by different doctors in reference to the kind of food that should be taken, under certain conditions, and that should be avoided, on the score of being non-digestible.

Many of the wise decisions on that point are from non-scientific sources. But they exercise an arbitrary influence over the minds of those who conceive it necessary to select a diet with express reference to its speedy, or rather easy, assimilation. And yet, gross mistakes are made, not through the false indications of science, but through ignorance of the first principles of chemical science.

For example, one recommends soft-boiled eggs; another, hard-boiled. Without being conscious of it, our likes or dislikes exert an arbitrary control over the judgment, and we think we are guided by scientific principles, when, in fact, we are managed by no principle at all in matters that purely concern the stomach.

Physicians differ exceedingly on the worn-out subject of diet. The various schools of medicine have their hobbies, while the representatives of each have their eccentric advocates.

Allopathics charge their patients as artillery officers load cannon, with all the gun will bear without bursting; therefore, ten grains of calomel, fortified with ten more of jalap, the practice of twenty years ago repeated, was the sheet-anchor of the old-fashioned practitioners.

Reforming homœopathics go to the other extreme. Struck with compassionate horror at the magnitude of incompatible compounds, they prescribe attenuated dilutions of something that can be neither smelt, tasted, nor felt. The one hundred and forty-ninth part of a grain, in forty gallons of water, is fearfully potent, administered by skilful hands.

Men of honor have never agreed in politics. It would be miraculous if there were no diversity of opinions in medicine. Each party is honestly impressed with the value of the dogmas they profess. Thus, inquiry is kept alive; otherwise there would be a stagnation of intellect, and another dark age. New and important truths are developed, in consequence of a difference of opinion among men equally honest and equally desirous of arriving at definite conclusions.

### Evanescence of Theories.

Theories have been repeatedly advanced from opposite directions touching the mooted question of what kind of food is best for human beings.

Civilization cannot settle the question. Savages give themselves no concern about it, devouring whatever is attainable that assuages the demands of hunger.

Notwithstanding the inculcations of physiological scholars, that certain modes of living tend to longevity, while others interfere with vital laws, and abridge the natural duration of life,—both savages and barbarians live as many years on the average, even less molested by the invasion of disease, than the most favored of mortals who fare sumptuously every day on viands that meet the approval of the soundest medical scrutineers.

We require a proper mixture of animal and vegetable food, it being of little consequence whether the first is roast beef, canvas-back ducks, sea slugs, roasted rattlesnake, boiled crabs, shark's fins, dried grasshoppers, fish, fowls, or turtle's eggs.

Some of all these usually considered disgusting, but largely consumed articles, actually nourish the body as completely as artistic dishes prepared according to the highest gastronomic authority, every one of them containing nutritious materials.

Science and civilization refine, but the empty stomach obeys an imperious law,—*eat or be eaten*,—making no apologies for dining on whatever satisfies the urgent demands of hunger.

The benefits derived from animal or vegetable food are to be measured by the results in respect to growth and reproduction.

### MECHANISM OF THE STOMACH.

A stomach is a receiving sac, into which food is taken, from which, by a series of extraordinary vital processes, materials are elaborated that enter into the composition of solids and fluids of which every living body is composed.

Every animal, small or large, except in the most rudimentary forms of life in particular families of infusoria, possesses a stomach, modified in structure to meet the peculiar conditions of each species. Some have two, some three, and the peaceable, patient ox has four, the food passing from one to the other before reaching the intestinal canal, where nutriment is separated from the useless matter with which it was united before digestion commenced.

All food requires a preliminary preparation before being swallowed. Thus, chewing, grinding, and lubricating it by being mixed with saliva,—a product of glands in the mouth and throat,—facilitates its descent down the œsophagus, and fits it for being more readily acted upon by the gastric juice.

The presence of food in the stomach stimulates its inner lining membrane to pour out a thin, bland fluid, which is a powerful solvent.

By alternate contractions and elongations of the fibres of that marvellously constructed organ, the mass is rolled to and fro, so that, being thoroughly mixed with the gastric juice, it is changed in appearance and consistence, preparatory to further vital processes.

Digestion is due largely to a succession of muscular movements commenced at the base of the tongue. One set of fibres takes up the action where those above leave the morsel, and thus it is propelled from point to point, till it falls, by its gravity, into the receiving-pouch, for such is the stomach in one of its functions, being quiescent till the cardiac orifice closes.

Teeth deserve a more extended consideration in this connection, than can be bestowed upon them at this stage of investigation of the laws of digestion.

As soon as they have ground down masses, and rendered them pulpy, soft, and easy for deglutition, they pass through uplifted arches at the top of the throat, not unlike a portcullis in their office. Fairly through, the gate closes, and next they are passed between two spongy bodies, the tonsils, the use of which is to oil them, as it were, to prevent friction or hindrance on the passage down the tube which leads to the stomach.

Finally, the circular and longitudinal muscular threads of which the œsophagus is constructed, contracting behind, urges morsels, assisted by gravity, till they fall into the membranous receptacle, where active chemical action is commenced.

## PROGRESS OF DIGESTION.

In a few hours, the food thus treated mechanically at first passes from the stomach through a narrow orifice, controlled by a sphincter muscle, which relaxes or spasmodically closes the orifice according to the sensation it receives from the approaching mass waiting to pass through the pylorus, into the upper portion of the duodenum, the first section of the intestinal tube, spoken of by old writers as a second stomach in man.

When a bit of bone, for example, has been accidentally swallowed,—a nail, a metallic button, a piece of money, or, indeed, anything that might produce irritation, or do violence in the intestines, it is not allowed to proceed, but is arrested as a prisoner in the stomach, where it is acted upon by the gastric juice till reduced to dimensions suitable for traversing the whole distance, nearly thirty feet, without injury to the delicate walls of the canal, then it is permitted to proceed.

The circular controlling muscle watching over the safety of parts beyond, is a vigilant sentinel that rarely ever fails of doing faithful duty.

Indigestible articles, or rather those which for a very long while resist the decomposing action of the gastric juice, move up to the pylorus in the mass waiting for exit through the gateway, but the never-sleeping watchdog—the sphincter muscle—detects the effort, and invariably drives it back.

Unless ejected by vomitation, an unwelcome traveller, urgent to go on the journey that he has commenced, may be thus retained for one or two years, and then be found in the stomach, if composed of elements on which the gastric solvent acts very slowly, or not at all.

SWALLOWING ARTICLES ACCIDENTALLY.

Pennies, thimbles, ivory and small glass balls, marbles and similar articles, the playthings of children, are often swallowed by them. When smaller than the ordinary diameter of the pylorus, such bodies are permitted to pass through unmolested, and they are soon voided without producing any disturbance or injury.

If, on the contrary, they are too large, they are detained till they have been so much reduced in size by the gastric secretion, as to pass with impunity.

Balls of hair are frequently found in the maws of cattle, when slaughtered, which must have been detained there a very considerable time, and which never could have been removed on account of their size, nor melted down to smaller dimensions, because their composition resisted the otherwise powerful chemical energy of the gastric juice.

They are of various dimensions, in cabinets from half an inch to four or five in diameter, and usually perfectly globular, as though they had been constantly rolling about to acquire that symmetrical form.

In the season of shedding their hair, cattle are in the habit of currying each other with their tongues. The surface of that flexible organ is covered with projecting eminences, called papillæ, which point towards the gullet. In raking off loose hair, it accumulates on them as it does on a currycomb. Not being able to dislodge such accumulations, and eject them from the mouth, they are swallowed. While detained in the first stomach, additions are made to the mass from time to time, which are matted on and felted there by mucous fluids, and, finally, the ball becomes not only large, but exceedingly compact, and hard as wood.

When a cud is raised to the mouth, those imprisoned balls, unquestionably, are also carried to the cardiac orifice, through which the cud ascends, but they are refused a passage. The same refusal is met at the other outlet towards the intestine. This, then, explains the origin and detention of such bodies in the stomach of ruminants.

### CHEMICAL POTENCY OF GASTRIC JUICE.

One of the most remarkable cases on medical record, demonstrating the irresistible solvent properties of the gastric juice—quite as intense in man, and nearly as concentrated as in sharks and serpents—occurred in Boston over fifty years ago, in the person of a sailor by the name of Cumings, who actually swallowed several pocket-knives. About one year after the event, two of the knives had entirely disappeared. The third was more than half gone when the patient died of gastritis.

Had the exact character of the case been understood, the surgeons and medical gentlemen in attendance at the hospital where Cumings had been admitted, not believing his constant assertion that he had penknives in his stomach, a course of tonic treatment might have been pursued that would have sustained him till Nature had completed the grand process of dissolving them, and thus relieving the poor sufferer, who was considered a lunatic.

When food arrives at the intestine from the stomach, it meets there with several peculiar secretions from small glands imbedded in its coats, each of which performs a specific chemical action on what is passing over the tract of their location.

## BILE OR GALL.

About twelve inches from the stomach, gall is poured into the moving mass, and various fluids from ducts opening into the interior of the intestine. A little lower, pancreatic fluid is introduced into the common avenue, which converts butter, fat, oils, etc., in an incredibly short time into an emulsion, which prepares them for digestion. Otherwise, without that particular fluid, those aliments would pass the whole length of the abdominal tube, and be ejected without having been essentially altered, or imparting any nutrition to the body.

## LACTEALS.

Still lower in that same membranous tube, minute orifices are discoverable in its walls, opening into it. Those are extremely numerous, and extend through the entire length, but are more aggregated into clusters in some places than others. Those are the mouths of lacteal vessels. There are millions of them scarcely larger than fine needles. The outer extremity running back, ultimately terminates in fleshy bodies, known as mesenteric glands. It is the office of those lacteal mouths to suck up, from the mass passing by, chyle,—a sort of milky-looking fluid, the product of digestion, which is carried directly into the mesenteric glands.

After remaining a little while there, probably mixing with a secretion peculiar to themselves, the fluid passes out through minute tubes on the opposite side, which finally empty their contents into a mealy-white tube lying on the side of the spine.

The mesenteric glands are way-stations, where the milky fluid, or chyle, undergoes chemical modifications before taking a departure for the thoracic duct, a reservoir into which the rich product of digested food, that which alone is nourishment, is conveyed.

Lying partly in front, but inclining to the left side, is a white ascending tube, under the name of thoracic duct, which finally makes a graceful curve, and enters into the great jugular vein at the root of the neck, at an angle formed by the junction of the subclavian vein from the arm with the jugular.

### WHERE THE CHYLE GOES.

A small, gentle flow of that milky fluid is constantly mixing with venous blood from the left arm and the brain, at the point described. From thence the new white fluid unites with blood that is on its way to the heart to be revivified, and loses its original color or whiteness.

Thus tracing the chyle from its origin, we ascertain the manner in which nature provides materials for sustaining and keeping in repair a living body.

### OXYGENATION.

Although material for making blood is thus explained mechanically, one further process must be completed to vitalize the mixture and fit it for the purposes of life.

Being carried to the right side of the heart, the auricle into which it is received contracts and forces it down through an orifice into the ventricle, a strong chamber.

That next contracts, it being a forcing-pump of prodigious power, and drives the new blood up through the pulmonary artery into the lungs.

When in the lungs, the blood thus driven in is distributed into unnumbered millions of fine tubes which ramify and spread round small air-cells. Next, we inhale air, which distends those cells into air-balloons. In the act of swelling with the inhaled air, the waiting blood imbibes from it oxygen, and then the lungs expel the air, thus deprived of an essential element, and, in expiration, throw off carbonic acid gas.

The blood is now vitalized and ready to fulfil its mission. For that purpose, being collected, it is again forced into the auricle of the left side of the heart. From thence it is forced into the ventricle of that side, and from thence driven into the aorta, a tube about three-quarters of an inch in diameter. That is ultimately subdivided into smaller and smaller arteries, by which the blood is freely distributed over and completely through every portion of the body, as already described on a preceding page.

### A Double Heart.

The right and left sides of the heart are quite independent of each other in function. There have been cases recorded where the two halves were separated at considerable distance from each other. Nature invariably pursues a system of economy in all her beautiful works, and this union of the heart of the lungs with the heart of the body is an illustration of the principle. By joining the two, less space was required, while muscular power was gained for both.

Such are some of the complicated processes on which life depends. A brittle thread, at best, is vitality, but without just so many cords, tubes, and tissues, there would be neither motion nor consciousness.

There is no difference in the anatomical appearance or structure of the digestive organs of males and females. They are precisely alike. The secretion of nutriment and its final diffusion in no respect differ in the two sexes. Their food, therefore, should be the same.

Women, in the higher social walks of society, oftener deprave their digestion than men, by subsisting on aliments too concentrated. This important fact is purposely repeated many times in this volume.

In the relation to which these remarks are applied, their food is not bulky enough, and consequently the alimentary canal is not as fully distended as it should be.

Some take food in too small quantities, for fear of obesity, and hence the abdominal region is gaunt and contracted, thereby compressing the hollow viscera too closely.

Those who by free exercise in open air have excellent health, also have an active digestion and a vigorous appetite. There is a better development of their frames; and both strength, beauty, energy of character, and those qualities which distinguish those who attain distinction, are due to perfect nutrition and freedom of body and mind.

The foregoing propositions may be considered trifling to those who have given no special thought to the philosophy of digestion. But the soundest, brightest, and most promising children are born of mothers who have a good digestion.

Feeble, sickly, peevish children, who live to become men and women, are always complaining and taking medicine. They had mothers from whom they inherited most of their physical, to say nothing of their moral and mental disabilities.

Numerous functional derangements, together with grave indispositions, are popularly charged to the liver. It is an organ uniformly supposed by those totally ignorant of its offices or construction, to have a controlling influence under circumstances where it probably has none at all.

Some physicians, especially those the least qualified by their anatomical acquirements to give a correct diagnosis, find it a convenient retreat for concealing their ignorance, to refer to

that organ as the seat of many morbid conditions, which cannot
be readily refuted if they happen to be wrong, on account of
its locality.

The liver is a gland of gigantic size, weighing in a woman
of medium stature about four pounds. Before birth it is vastly
larger and wholly disproportioned to other organs in the abdominal cavity, as they appear in adults.

A reason why it necessarily has such dimensions is in consequence of having nearly all the circulating blood from a
maternal source sent directly to it. At birth, with the first
breath of the infant, one half the blood that went to the liver
before is instantly diverted from it by the closing of a valve in
the middle of the heart.

In consequence of being thus suddenly deprived of so much
vitalizing fluid, the liver hardly maintains its volume. Certain
it is, it remains stationary in size for a long while. In the
meantime, other parts which were somewhat rudimentary, as it
were, or imperfectly developed, grow into their predestined
proportions and assume more active labors.

From blood sent into the liver, gall, that intensely bitter
fluid, is secreted. One of the specific uses of the liver is to
elaborate that extraordinary product from venous blood. Arteries convey florid, vitalized blood to the intestines and digestive
apparatus, where it leaves its vitalizing influence. When that
is extracted, the remainder flows through another set of vessels,
veins, which carry it to the lungs, to be recharged with oxygen
from inhaled atmospheric air. On its way there it is compelled
to pass through the liver, and from it certain vessels take out
of it bile, and, as we shall learn in the sequel, some other
products.

## INTRICATE MECHANISM.

No mechanism, on the whole, is more intricate than the network of tubes by which bile is separated from the passing current of venous blood. When detached or drawn aside by itself, a transfer of it to the gall-bladder, where it is stored for after occasions, is one of the great curiosities of animal construction.

Physiologists, with all their ingenuity and indomitable perseverance, have not yet definitely settled the question of the use of bile in the economy. That it is of importance in digestion can hardly be doubted, and yet there are more theories extant than facts to show where it goes, or what it is for.

Bilious affections, bilious stomachs, a bilious habit, and such like expressions, are flippantly banded about by medical practitioners as they are by persons who learn them as parrots do from hearing repetitions of the same phrases, without attaching any meaning to the words. It is an evidence of ignorance rather than scientific attainment, when guessing passes for profound pathological acquirements.

Too much is charged to the poor liver, and tons of pills and useless prescriptions are directed to the correction of faults it never had—to the cure of diseases in which it had no agency.

Regarded by non-professional persons as performing offices which it does not perform, their deductions are, of course, as crude as those who pretend to more knowledge without being a whit wiser. Bile is considered a terrible foe to health in common parlance, a disorganizing bugbear, a maker of melancholy, a breeder of low spirits, jaundice, and a host of other misfortunes that beset mankind.

Allusion has already been made to the sugar-making services of the liver, coupled with observations on its complicated
functions before and after birth.   It being a comparatively
recent discovery that man and all the lower families of terres-
trial animals carry within their bodies a sugar-mill, we cannot
pass over the natural provision for meeting the demands of
organic life, without dwelling particularly on that remarkable
function, on different pages of this work.

### DEMAND FOR SUGAR AND BILE.

Sugar must be provided from some source.   If it does not
exist in sufficient abundance in the food of each day, the
deficiency is supplied by the liver,

Attached to its largest lobe, lying underside of the diaphragm to which it is attached, is a small bag, about the size
and form of a small pear, into which bile is stored for future
use.   A slender duct leads from it to the first portion of the
small intestine some twelve or more inches from the stomach.
In the process of digestion the bile flows into the upper end of
the intestinal tube, and, undoubtedly, there performs an active
part in chemically preparing the passing food for yielding up
its nutritious elements; but what becomes of it afterwards has
not yet been positively ascertained.

Carnivorous animals secrete more bile than graminivorous;
and ferocious fishes, as sharks, torpedoes, wolf-fish, etc., require
far more than social dwellers of the sea.

Admitted to be indispensable to perfect digestion, how it
acts, or what becomes of the quantities secreted, since it does
not pass off with waste materials in the ordinary manner, very
much exercises the inquiring minds of physiologists.

That noble servant of man, the horse, feeding exclusively
on vegetable food, as those animals do which chew the cud, has

no gall bladder. His liver is of ordinary appearance exteriorly. If bile is secreted in the horse's liver, where are the excretory ducts that conduct it to the food?. But that an organ of such magnitude and weight, occupying so much room, has no service to perform after birth, is hardly probable. Naturalists have the mortification to acknowledge the impossibility, at present, of explaining its true function in the horse.

Ignorant as we are, and humiliating as is the confession, that many guess at much they do not understand, the diseases of the liver are of a character to perplex and baffle the most experienced physicians.

In certain climates it becomes indurated, enlarges enormously, and besides, scirrhosity, abscesses, and ulcerations are common in all climates, as a penalty for violating sanitary laws, which can never be pursued for any great length of time without a fearful constitutional reckoning.

Malarious influences emanate from the ground in warm, moist regions, where vegetable decomposition fills the air with something neither seen nor tasted, but which, nevertheless, when inhaled, produces extraordinary disturbance in the liver of man. Thus, fever and ague are derived from that source, while another condition of the atmosphere in the East Indies gives rise to various enlargements and hardness, which defy the ordinary resources of medicine.

Authors have not sufficiently investigated the effects of certain kinds of food in the production of anomalous disorders of that viscus. That the profligate use of curry in the East Indies—a fiery hot powder made of red pepper, mustard, turmeric, and perhaps a dozen other ingredients, which would excoriate the skin, externally applied, about as quickly as a burning coal,—taken into the stomach at every meal for years in succession, must, in the nature of things, derange not only the stomach, but associated organs. So it may be admitted, inasmuch as curry-eaters have indurated livers, that peculiar appetizing compound has some agency on the organ.

Whiskey, rum, or, indeed, any of the fiery strong liquor disgracefully in request in this whiskey-smitten nation, acts banefully on the liver. Those who keep themselves stimulated by the needless use of distilled spirits, must break down under its undermining tendency. Medicine furnishes no cure for an enlarged or schirrus liver.

When it becomes diseased in any way, then there is a failure to perform the office for which it was mainly designed, and, consequently, the whole body quickly betrays its need of something it formerly had, in a yellowish, or rather, a deadly hue of the skin, loss of flesh and strength, and waning health.

Next, that which is required is a sufficiency of saccharine matter, from which are elaborated, by vital processes, elements to be distributed for the benefit of the whole body, and perhaps, too, for the mind.

The liver, in short, manufactures sugar. It is not exactly sugar of the shops in appearance, but a sweetish paste, that takes the name of glucose.

The mass of the liver appears to be made up, in bulk, of an immense congeries of arteries, veins, nerves, lymphatics, biletubes, ligaments, and a semi-elastic tissue, which serves as a bed to keep all those different parts from interfering with each other.

When this natural sugar-mill turns off more sugar than the system requires, it is recognized as a disease known as *diabetes*. Nature has but one convenient way of carrying off the excess, and that is by dissolving, and floating it away to the kidneys. Those organs separate the sugar from the blood in which it arrives, and forwards it to the bladder to be voided.   By boiling the urine, the sugar may be collected, very much resembling ordinary brown sugar.

## Diseased Livers.

Severely as the liver suffers from over-excitation by drinking ardent spirits, an instructive article might be written on the unnecessary medication to which the whole system is subjected by the mistakes of physicians, who blindly pursue a course of practice based on a theoretical condition of the liver, for which the poor stomach is intolerably dosed.   There is no more direct means of reaching the liver in any of the morbid conditions to which it is predisposed from climate, abuse, or from dissipated habits, than through the circuitous route of the circulation.

Mercury was formerly prescribed immoderately, on the supposition that the liver was answerable for at least half the ills to which humanity is incident.   Salivations, ulcerated tonsils, loose teeth, inflamed gums, and even caries of the bones were the result of that one-idea practice now obsolete.   But the liver was too frequently the focus to which nauseous preparations were directed, when it was, perhaps, in no way involved.

It is impossible for any medicine to reach the liver directly. There is no tube or avenue opening between the stomach and liver.   Therefore, it is ridiculous to suppose the latter can be acted upon in any other manner than through the blood.

Some persons, more distinguished for general intelligence than their knowledge of anatomy, speak of ulcers, or abscesses of the liver, which discharge into the stomach.   That is positively impossible, unless an opening has been ulcerated through various tissues, and lastly, through the walls of the stomach, before any such imaginary communication can be established.

Hastily fattened cattle, stuffed with rich food faster than it can be appropriately digested, or fed on warm slops, become singularly disordered in their liver. Red spots, ragged, ulcerated patches on the upper surface, and enlargement, evidently show that properties may be introduced into the circulation, which, on arriving at the liver, are arrested, and stopping there, throw the organ into a morbid state of action.

Hypertrophy, induration, and abscesses are conditions of the liver in men who have no mercy on themselves by excessive indulgence in strong liquors. Women rarely have diseased livers. Happily, they have a nicer sense of propriety. Their livers seldom become disorganized, or suffer from those hepatic woes that beset tipplers.

But women induce hepatic difficulties by a custom in dress, indicated by a yellowish, tallowish complexion, usually associated with a depraved appetite.

Tight-lacing compresses the right lobe, lying just behind the short ribs; if the waist is closely girded, that part of the organ is pressed into close quarters, which must interfere with a free circulation of the various fluids which it secretes, independently of arterial, venous, and biliary currents.

If the bile is impeded in its progress to the gall-bladder, or from thence into the intestines, in consequence of ligating the waist, very serious consequences are liable to follow.

Here is found an explanation of an often asked question, Why young ladies are so frequently tinged with yellow, accompanied by indigestion? The bile is obstructed by compression of the liver, by waists of dresses and belts, and being taken back into the system by absorbments, is diffused over the body, and escapes through the skin. Jaundice is simply that condition,—the bile not flowing off through the pipes in which it should go, owing either to exterior mechanical compression, gall-stones, or an inflammation which closes them.

A celebrated manufacturer of corsets, having satisfied herself that women will wear them—which is admitting there is no necessity for that kind of abdominal support—has invented a substitute. It may be worn with comfort, as it neither compresses the chest, ribs, nor the sternum. Her object is simply to hold up the bowels, so that they cannot be forced down upon the pelvic viscera. Having the confidence of physicians, the inventress has extensive patronage, because the contrivance actually relieves the pelvic organs from invasions which ordinary stays produce.

# CHAPTER XVIII.

## Their Growth.

Men Taller than Women—Male Animals—Physical Aspect—Length of Lower Extremities—Osseous Development—Suspension of Growth—Inner Capacity of the Chest, Broad, Narrow—Short Necks.

Every circumstance in the history of an individual life, in a physical aspect, must be influenced by laws which govern all organized bodies. Even inorganic forms are regulated by fixed laws also, since there is nothing transpiring by chance.

There is a law of limitation in the growth of men and women, operating infallibly in the formation of each and all tissues, by which proportions are established.

Men ordinarily are taller than women, and stronger. Males of all orders are usually superior in size, and muscular force in them is also proportionately superior to that of females of their kindred. Such is particularly the case with quadrupeds and birds. They are more beautiful, too,—more imposing in their physique and bearing. Females are smaller, and destitute of those markings or colorings which are distinguishing beauties, including manes, fringed limbs, brilliant feathers, and other exterior appointments that give character to the males.

Woman, however, transcends in beauty of form, facial expression, and in the impression she makes on the spectator.

Among reptiles, usually, the female is the largest. A law of positive necessity operates in favor of that oversize above the male. The enormous number of eggs some of them extrude, or the number of young incubated within their own bodies, requires room for the expansion of oviducts in which they are carried.

Thus there are oviparous and viviparous reptiles. Some void their eggs to be incubated by the solar rays, while others have them hatched in the abdominal cavity.

Birds, being of a higher type, have their eggs developed so that one is voided daily, or once in two or three days, ripening so orderly and rapidly too, that a larger pouch is not needed. If their eggs all matured at once, as in a turtle, a fish, or in thousands of insects, in parcels, which are extruded at intervals of one or two weeks, the bulk of the eggs laid in twenty days would equal, if not exceed, in bulk the body from which they were extruded.

Some tribes of fishes have amazing fecundity, actually producing millions of eggs in a single season. Were they brought together, their combined weight would exceed the weight of the individual in which they were formed by twenty-fold.

The rapidity of development of some insect eggs in a single day, from mere specs scarcely discernible, into fully distended globes almost as large as peas, illustrates in another form the extreme activity of vital force when aided by light, heat, and moisture.

## TALL OR SHORT.

Why a man ceases to grow taller on reaching six feet, six feet four inches or more, or why growth is ever arrested in the process of osseous elongation, is quite beyond the ken of modern philosophy. Theories prove nothing, while facts cannot be jostled out of sight. Speculations on this point, therefore, are to no purpose.

Admitting that men rarely exceed six feet in any country, if a few happen to exceed that ordinary standard of limitation, they are called giants. Why women rarely reach the same measure, is quite as difficult to explain as the other proposition. A difference in height depends almost entirely on the length of the lower extremities in both sexes.

From the crown of the head to the ischiatic knobs,—two points on which we sit,—there is not much variation in the measure. Males and females have an equal number of bones, and the distance between these two starting-points is about the same. Below, however, the length of the thigh-bones determines the stature of the individual.

Some singular anomalies are noticed occasionally, which seem at first view to contradict a received opinion respecting the laws of growth.

George W. Crawford, of Sciota county, Ohio, fifteen years of age, in the autumn of 1869, was six feet and one inch tall, measuring around his shoulders three feet eleven inches; around his hips, forty-two inches; around the chest, forty-one inches; and he weighed two hundred and eight pounds.

Benjamin F. Kiplinger, of Rush county, Indiana, about the same period, who was fifteen years old September 20, 1869, stood six feet eight inches, measuring around his shoulders fifty-seven inches, forty-six around his chest, forty-six around the hips, and weighed two hundred and thirty-five pounds,—wearing number twelve shoes!

Seated at table, on the same level, men and women, taken indifferently, appear to be about equal in height, there being only a slight deviation from a horizontal line passing above their heads. On rising, some are exceedingly tall and others remarkably short. The difference is found in the femoral bones. From the knee to the instep, the tibia and fibula, or leg-bones, are correspondingly short also, to conform to proportions above the articulations.

At birth the lower limbs are very short and small, quite disproportioned to the scale of development of the upper extremities, which is explained by the well-known fact that they receive but a limited amount of blood while in utero. Immediately after birth, blood which circulated in the placenta, diverted from the iliac arteries, is then sent into the legs. But they seldom attain in females the length of the lower extremities of males, even when the nutrition is increased by an increased flow of blood. Hence, women are generally below the stature of adult males. Exceptions to the rule are considered anomalies.

Blood is circulated very nearly alike in both sexes, but the extension of bones is more actively carried on in boys than in girls, in bones below the pelvis.

This law of osseous development presents matter for consideration in regard to life-insurance investigations. *Physical signs of longevity in man* was a prize essay a few years since, published by a Life Insurance office of New York, abounding in very curious facts not very generally known in relation to life limitation. Some of them were as follows:

First,—Brothers and sisters of the same parentage, reared under precisely the same circumstances as regards food, clothing, ventilation of apartments, etc., have different statures when they arrive at adult age. Yet at birth, and through the developing periods of childhood and adolescence, they were apparently influenced, physically, precisely alike.

Unquestionably, therefore, there are causes operating disadvantageously, at times, for the growth of parts, if not of the whole body. In dwarfs, the deposition of ossific material stops suddenly. It may happen soon after birth, or at any period between the second and third year. Occasionally the process of growth ceases in a single limb, or it may in both so exactly at the same time as to leave them of the same length. From some unknown cause, essential elements cease to be any longer deposited.

While there is a progressive development in the system, and all the mechanism is being enlarged in volume and perfected, there is intense activity. By and by, however, the law of limitation puts a stop to those long-continued internal operations.

Ossification is then completed, the muscles are full and strong. The future secretion and deposition of lime and other earthly components of bones, instead of being gathered in such abundance as formerly from food, are only just enough to keep those solid parts in repair.

### Strength of Bones and their Decay.

An impression is entertained that bones of tall persons are more easily fractured than those of short people. Cylindrical bones, as the thigh and arm, when particularly long, are less in diameter than the same bones in those of short stature.

As individuals advance in age, gelatine—the mortar that holds the bony particles together like bricks in an edifice—is secreted less actively, and its adhesive properties are also enfeebled. Finally, the quantity is so much diminished, the bones of aged persons are easily broken. A sparseness of that natural glue explains why their fractured bones unite slowly, or sometimes not at all.

If, as some surgeons suggest, broken bones of short patients unite quicker than those of tall ones, all other circumstances being equal in respect to age, attentions, etc., it must be due to a more rapid circulation in the first, in whom the pulsations are quickest and most energetic.

It is a fact that vital force is strongest in short people. The blood has not so far to move, and there is less retardation of the current from friction, admitting that curves and short angles in arteries and veins offer some resistance.

A general impression is entertained among close observers, that longevity appertains to persons rather under size than to the tall.

A broad, full chest does not always belong to a tall man or woman. On the contrary, those under size are rarely fragile in form, or narrow across the thorax.

When the inner capacity of the chest admits of a perfectly full inflation of the lungs, the prospect of life is greater than in a constricted cavity where the organs cannot have play enough to oxygenate the volume of blood sent to them.

## SHORT WOMEN.

When solidification of the leg bones progresses slowly, there is commonly an active ossification taking place in the spinal column. Harmonious architectural proportions are not maintained in women as in men. There are more short females than males. Perhaps it may be there is a predominance of short men in a thousand, but whether tall or short, the scale of proportions is superior in the tall.

Among a thousand females of all conditions of life, the short immeasurably outnumber the tall—the upper parts of their bodies being generally better developed than the lower, which are not in exact proportion with the scale above the pelvic arch.

A lady may have a finely-developed chest, a round full bust, well-set shoulders, and a beautiful neck, while the thigh bones are so very imperfectly developed that she is disproportionately short.

Every internal organ, embracing the entire contents of the thorax, abdomen, and pelvis, are quite as large and perfect in function as in those ladies who are tall. The only anatomical difference is to be found in the length of the bones in the inferior extremities.

A lady distinguished for a particularly long neck, swan-like in gracefulness, may be considered to have an imperfect chest, and, therefore, her life expectation is not as good as that of one of the same age and physical condition whose neck is an inch shorter.

An explanation of this law of probability is found in the osseous structure. All men and women have twenty-four bones in the vertebral column, seven of which are usually in the neck. Those twenty-four blocks, which, collectively, are called the spine, are singularly locked together to prevent them from sliding out of place.

Occasionally an anomaly is recognized in the distribution of these bones. There should be always seven in the neck, twelve in the back, and five in the loins. But when the neck is unusually long, it has eight blocks. That takes one from the dorsal range, leaving only eleven in the back.

That circumstance necessarily makes the chest just the depth of the missing bone smaller, in its vertical direction, than it would have been had it remained where it is usually to be found.

### CAPACITY OF THE FEMALE CHEST.

The lungs and heart, as a natural consequence, are compelled to act in a smaller cavity. That being the actual condition, those vital organs, on which the preservation of life depends, are cramped, and their expansion limited in the performance of their functions.

Thus, if the lungs have not room enough for full inflation, nor the heart for its diastole, the consequences are unfavorable for long life.

Here, then, is a plain mechanical demonstration of the anomaly of a long neck, and the consequences resulting from diminishing the capacity of the chest.

When the neck is remarkably short, it may have seven bones in its composition, but they may be so thin as to be a deviation from the type which nature in most cases prescribes.

Thus, the chest may be full and broad, while the physician recognizes in that kind of short neck a tendency to apoplexy. Irregularities or excesses of any kind, including sudden excite ments, pain, stimulants taken into the stomach, excessive paroxysms of rage, hatred, love, or joy, drive blood into the brain faster than the veins conduct it away, and sudden death ensues.

With a short neck and large chest the heart acts with great energy, forcing blood into the brain and deranging it, on account of the inability of the veins to carry it away fast enough.   This is apoplexy.

# CHAPTER XIX.

## Their Eyes.

Force of Ocular Expression—Wearing Glasses—Desiring to Appear Near-lighted—Fashionable to have Defective Vision—Abuse of the Organ—Eyesight of Animals in General—Do without Glasses if Possible.

For brilliancy, no gems compare with the eyes of a beautiful woman.

Examples are unnecessary for establishing the truth of this declaration. There is a fascination, a bewildering influence in a pair of bright eyes that moves and, indeed, electrifies the roughest specimens of manhood with undefined emotions.

Fine eyes are potent engines. When the features are symmetrically moulded, eyes of some hues are irresistibly powerful. Set off advantageously by long silken lashes, a sweet expression is the highest type of female loveliness.

Men cannot explain, even to themselves, how or what it is that moves them so mysteriously in coming into the presence of a handsome woman. It is admitted there is an irresistible force set in motion, but in what manner it takes such hold is not of easy explanation.

That magnetism—an unseen agent—is the instrumentality with which women are made more potent than the strongest men, cannot be questioned. It is more than an equivalent for large bones and elephantine muscles.

Men brave tempests, dare enemies in bloody combats, looking destruction in the face with unflinching energy. Woman shrinks back in timid consciousness of being unable to battle physically for her rights. In the very posture she assumes, the expression she exhibits, and the delicacy of her organization, she is more than a match for giants when they offer violence.

When an incensed woman fixes a withering glance on a wretch who threatens to do her wrong, or calls her honor in question, the weight of her scorn is unbearable. Such a villain suddenly cowers beneath her searching indignation, and wilts away from her heroic presence. //

Regardless of color, the eyes singularly harmonize with the features. Complexion and general corporeal expression is a study gifted artists have not yet mastered, although they have been pursuing their investigations since the days of Apelles.

## VISION.

Perfect vision is marred, and, indeed, the eyes that were perfect, and would have remained so through the ordinary circumstances of life, are seriously injured now-a-days by the caprice of fashion.

### INFLAMED EYES.

Inflammation of the conjunctiva, the first membrane over the front of the globe, delicately thin and transparent, is kept slightly inflamed by too much light. The pupil—a round window through which light reaches the posterior wall of the eye—cannot escape injury, if the outer membrane becomes either thickened or clouded. Both of those conditions may be induced by applying preparations with a view to making the eyes more brilliant. It is a weakness of a very extensive class of ladies, who, in their desire to make them piercing, or, as they imagine, more captivating, cannot be convinced those preparations they use are lamentably injurious. An irritant that inflames any surface extends its influence beyond where it is applied. Anything that directly offends the irritable anterior surface of the eye, instantly brings a flood of tears to wash it away. Streaks of blood are simply an engorgement of minute vessels, which, unmolested, are invisible behind the conjunctival membrane.

Inflammations thus exhibited conclusively prove there has been some wrong-doing, or incidental exposure to causes which produce that condition.

When an inflammation is established, and the vessels under the conjunctival membrane lying on the selerotica, or white of the eye, become strongly defined, if not subdued, they may shoot across the pupil, forming a veil that would obstruct the passage of light.

When there is a sensation under the lids like particles of sand, it indicates the development of projecting fleshy granules on the under surface, which chafe, and still further increase inflammation by the movements of the eye. The friction is intolerably painful in some cases, accompanied by an intolerance of light. Improper applications to the organs, in the form of washes or unguents, keep up a continued irritation that may result in the production of granulation or other equally severe afflictions.

Some persons are predisposed to a preternatural irritability of the margins of the eyelids. They have a red, inflamed appearance, generally aggravated by a sudden cold, a particularly strong light, or exposure to winds laden with dust.

A peculiar ferreted appearance of the lids, which is a chronic inflammation of their most exposed mucous surface, is attended by another inconvenience that may degenerate into a formidable malady, if too long neglected. It is a gluing together of the edges of the upper and lower lids by the flow during sleep of an adhesive secretion, slowly soluble in cold water. Tepid water separates them pretty readily.

A continued use of cosmetics, apparently perfectly harmless, not unfrequently do great injury to the eyes of ladies who indulge in that reprehensible practice of attempting to improve upon Nature.

Eyes are constructed upon philosophical principles, so perfect with reference to the laws of light, that they cannot be tampered with, nor readjusted easily when once disordered. The refractive power of the lens may be altered by violence inflicted on the exterior of the globe.

So much of our knowledge, happiness, and every-day comfort depends on a sound, perfect condition of our eyes, we cannot be too choice of them. They are too precious to be jeopardized under the treatment of ignorant, self-announced oculists.

## WEARING GLASSES.

Many charming faces are completely bereft of the expression they would have had, unmolested by the silly desire of otherwise sensible ladies, for wearing glasses. An unaccountable disposition to have it supposed that they have defective vision, is another strange phase in the vagaries of fashion. To be near-sighted is a coveted grace.

In some departments of elevated society, nothing is more common than to see young ladies harnessed in spectacles, or peering through an eyeglass at their familiar acquaintances on the side-walk, as though it were extremely difficult to see them at all.

None but fops or idle pretenders of both sexes, who ape the artificial manners of some polar star in fashionable circles, think of making themselves ridiculous in that particular way.

It is conclusive evidence of their vanity and mental weakness. An eyeglass dangling from a splendid chain is a coveted ornament for a drawing-room. To be squinting through it at wall-pictures, or closely examining an object that a blind man might almost see, by those who have no imperfection of vision, is a common folly.

Everything, near or distant, must be scrutinized through an eyeglass. Not because they cannot see, but simply because it is extremely genteel to be purblind.

To gaze with profound attention through an eyeglass at a horse passing the window, with an avowed inability to determine what creature it may be by the unassisted eye, is an immense recommendation, indicative of polished manners. If a lady is ingenious in striking attitudes at the same moment, she may consider herself a queen of fashion.

No vulgarity is rated lower in the tablet of exquisite refinement, than having good sound eyes. Examining those to whom one has an introduction, with an eyeglass, as an entomologist would scrutinize a bug under a microscope, passes for extreme refinement.

Young misses, fresh from a boarding-school, are in ecstasies when they first have possession of an eyeglass set in a chased gold rim. They then cannot see those they do not wish to recognize—which is a decided step in their education.

### PROGRESS OF GENTILITY.

On the whole, it is deplorable that civilization delights in blindness. Possibly a sentiment prevails that one can see enough with half an eye. But this, absurd as it is, is associated with another equally ridiculous habit, that has even got possession also of men of the no-brain order. To lisp divinely, and be in poor health, is the climax of perfection in the constitution of a modern lady of unexceptionable social position. It gives a finishing perfection to a belle of the period.

These are follies that amuse people of sense for a while; but it is, nevertheless, lamentable that folly should have such prominent ascendency where genuine good-breeding and worth of character are at a discount.

Those who cannot afford to be blind voluntarily, like those who articulate their words distinctly, have no influence where near-sightedness and lisping are the criteria of social excellence.

Near-sightedness is most appreciated in circles distinguished for opulence. In the country, remote from the baneful influences and innovations of fashionable folly, the ladies have eyes keen enough to discriminate between affectation and malformations.

A real necessity for glasses appertains to advanced age, but rarely as necessary as those who have them to sell would have the world believe.

There is another unrebuked exhibition of vanity or self-esteem—it is difficult to determine which—viz., having portraits and photographs saddled with lunettes at the expense of a silly unmeaning expression. Artists dread them, knowing by experience of the impossibility of giving any character to the picture of a face marred by bows and glasses.

Portraits of men and women with strongly moulded features, full, animated eyes, in harmony with their other physiognomical attributes, are deprived of an essential part of their force of expression when painted in spectacles.

It is quite surprising with what tenacity some young, newly-fledged clergymen cling to glasses, whose eyes never had a

defect in them, on the presumption, it is theoretically presumed, that an audience associate with such toggery, profound
scholarship, and deep theological explorations in the dust
of ages.

No orator who moves the multitude by the power of his
eloquence, wears glasses. To touch the hearts with fitting
words, to arouse the deepest feelings of sympathy, or excite
ferocious indignation by a recital of real or imaginary wrongs,
the full, unshackled face of the speaker must be seen. Sentences that roll along the aisles like avalanches from the lofty
summits of mighty mountains, would lose their effect if enun-
ciated in the dark. An orator must not only be seen as well as
heard, to accomplish the highest results of his burning language, but his face, and particularly his eyes, must not have
their electrical energy intercepted by non-conducting glasses.

Each one of the special organs of sense is a faithful sentinel
till the hour of death, if it has not been impaired,—even
beyond a hundred years, in vast numbers of instances.

Taste and feeling rarely ever flag in a prolonged longevity.
When three other senses are destroyed, there is consciousness.

Through the instrumentality of nerves, the mind receives
intelligence of impressions, of whatever kind or character.

Vision ought not to give out till the lamp of life goes out in
old age. Were we to treat our eyes with as much tenderness as
they deserve, we should have distinct vision till the hour of
death, at the most advanced period of human life. Our eyesight would be nearly as perfect when we have reached seventy
years, as when we were young, were it not for the abuse of
them by intense light, gas-jets, and the fatigue to which they
are subjected by reading small type-books, and continuing the
labor too long at a time.

Wild animals have perfect vision as long as they have ability to forage for food. Birds, too, have distinct vision till they die of old age. A goose lives to upwards of eighty years in a state of domestication, with no failure of vision. Probably, reptiles and fishes also have perfect and accurately distinct vision at all periods through their long lives. If whales reach a thousand years, and sharks an extended longevity, their vision is, unquestionably, perfect and unimpaired all the days allotted them.

Tortoises have been repeatedly found with dates inscribed on their shells, indicating almost a century from the date of the marking, and they may have been ancient settlers when those dates were inscribed; yet their eyesight was keen enough for perceiving an enemy, or discovering appropriate nourishment.

Man, alone, has defective vision prematurely, and usually from neglect or over-working his eyes. Domesticated dogs, cattle and horses in the service of men, are subjected, to considerable extent, to conditions of exposure, which impair our own sight. A dog reposing in the corner, occasionally gazes into a blazing fire. Horses and cattle are approached with caution in the stall, or placed where artificial light acts directly upon their eyes. When, under the guidance of their own instincts, they retire, as the fowls go to roost, with the approach of night, and open their eyes early, as the sun gradually rises, so that no sudden glare impinges to their injury.

All animals avoid light, after evening shades set in, unless compelled to change their habits. That is the secret of their excellent and distinct powers of perception.

Were we to do as they do, we should have no complaint to make of waning vision.

## CHANGING AXIS OF VISION.

The convexity of the eye undoubtedly varies so that scarcely any two persons have the same curve, and hence the focal distance of distinct vision must necessarily vary. One sees accurately at ten inches, another at twelve or fourteen, and another at eighteen or twenty inches from the eye. The scale of distance varies exceedingly in that respect, in the small number of a dozen persons.

In examining the moon, it rarely happens that twenty ladies and gentlemen agree in their estimate of its apparent diameter. To one or two it may seem about two feet across its face. Others are quite sure it is all of a yard, and possibly, it scarcely appears much larger than Venus to another.

By practice,—beginning, for example, in early childhood with the alphabet, and gradually learning to read with facility —the visual organs are trained so systematically, that we usually all have a focal point of clear and distinct vision at the ordinary distance at which a book is held for reading. Our eyes are systematically educated, as our legs for walking, or our tongues for articulating words.

Beginning in childhood, we insensibly instruct our organs of sense and our muscles, and finally they all harmonize at last; and judgment, which distinguishes man above all the races below him, is perfect or defective, according to the development of all the powers which belong to his physical organization. Perpetual repetitions of the movement give to each and all those parts controlled by our volition, the perfection which they may attain.

In early youth there may be some rigidity of the cornea, which does not readily yield to the training. If the curvature is too prominent for seeing at ordinary distances, most convenient for looking at a page, and practice in trying to see at that convenient distance is not successful, there is near-sightedness.

## NEAR-SIGHTEDNESS.

When just that condition has been ascertained, parents and the near-sighted child too hastily resort to concave glasses. If they would resolutely insist upon an unremitting effort to do without them, their eyes would gradually accommodate themselves to the task imposed, and vastly improve.

Prematurely putting on glasses arrests the progress of adaptation, which would very certainly take place, although in every instance it might not become entirely satisfactory. The experiment, however, is worth trying.

Avoid glasses as long as possible, whether short or long sighted, and thus allow the instrument to adjust itself to circumstances. The eyes of all land-seeing animals are constructed upon the same principle as our own. Light is admitted into the back region through the pupil, and there produces the same impression as it does in men and women. There is very little, if any, real difference discoverable in the anatomical structure in day-seeing eyes. But those of wild animals wear longer without becoming impaired, than the eyes of domesticated animals or man, simply because they act in conformity to natural laws. Daylight, while they are ranging over fields carpeted in green, or forests in which dazzling rays cannot act directly upon them, favors them exceedingly.

Oculists and spectacle-manufacturers are reluctant to admit the existence of this law of ocular adaptation, which is quite as readily demonstrated as many other problems of less importance.

## VISION OF AGE.

Before spectacles were invented, there is good reason for believing that people had better eyesight than since. Historians speak of the blind, but nowhere is there a lamentation over the waning vision of old age as in modern times.

When, in consequence of advancing age, glasses are resorted to, they must afterwards be continued. The eye seems to lose its power of adaptation to varying circumstances, whenever artificial aid is provided. In other words, if glasses are prematurely worn—and they generally are prematurely put on, according to our theory—they cannot afterwards be laid aside without inconvenience.

When the time comes, as it does in the life of each of us, that the eye is less prominent than it was in youth, vision is less distinct than before, and we meet that flattening of the cornea by convex glasses, which apparently enlarges the letters of a book, and therefore they are more distinct.

That is precisely the period to resist the aid of glasses. Have patience, and regularly exercise the eyes to reading at the same convenient distance they were formerly used, and they will, after a while, return to their primitive convexity.

Will-force produces extraordinary results. Even pulsations of the heart have been suspended by it, and the organ again set in motion by the same agency.. It is even claimed that it is possible to exert that mysterious nervous energy, so as to positively control the volitions of others.

With the approach of old age, there is a gradual relaxation of all the tissues of the body. Those of the eye lose their former tension, and the secretion and removal of the fluids within the globe on which the refraction of light depends, as also the chromatic perfection of the picture on the retinal canvas, is sluggishly performed. But if we urge them to the performance of their office, they begin to receive more vital influence, and readapt themselves to the work demanded of them. In short, the determination and persistency of effort may be crowned with success.

Without burdening these pages with narratives of eminent success by pursuing this course, it is quite sufficient to say that failures would be few in making the experiment, if those who are making it would on no account deviate from the directions proposed.

After weeks of hope, without apparent amelioration, two-thirds of those who may have commenced with a strong resolution to be thorough in their attempt at visual restoration, become impatient and fly to glasses, and then doubt the possibility of seeing without them in after periods of life.

Professed oculists are the bitterest foes with which the advocate for having nature consulted first, comes in contact. To a man, they recommend glasses number one, two, three, and so on, with a farago of nonsensical reasons for favoring the eye when it requires no such aid.

## DURATION OF VISION.

Our eyes were designed to last as long as the sense of hearing, taste, or our fingers and toes; and they would, were they not culpably abused and overworked by the customs and habits.

Blue eyes are thought best adapted for all climates. Black predominates in tropical and semi-tropical countries. The farther north, the lighter the blue shade; and it is among the blue-eyed that the fewest glasses are worn, according to the observation of travellers. Such eyes possess qualities for a more distinct vision, all other things being equal.

Black eyes are lustrous, and carry with them an intensity of facial expression superior to gray, or any of the lighter shades of color. Black, hazel, etc., if not quite as liable to cataracts, or less formidable opacities, fail earlier than blue, subjected to the same treatment of brilliantly-illuminated rooms, bright blazing firelight, gas jets, and similar sources of injury.

There are beautiful blue-eyed ladies with blonde hair. The iris and hair generally are alike in color. When eyes are too light-colored to be sparkling, the hair is ordinarily yellow, and the brows thin and colorless.

With heavy dark eyebrows and black eyes, the expression is strong, and not unfrequently imposing.

Pretty female faces, with small eyes, cannot be roused into a look of majesty, although capable of inspiring poetical sentiment.

A tragic face must have full black eyes. A tragedian with light eyes must rely more upon costume for success than on his features. A grand, imposing actor, male or female, must either possess dark eyes, or divert the attention of the theatre by artificial devices—voice, dress, and gesture being the handiest instrumentalities.

There are actors whose faces alone, without the utterance of a single word, set an audience in a roar of laughter. And there are also players of another grade, who command a spontaneous burst of applause the moment they come in sight upon the stage, before they have uttered a word.

With a continuance of the present fashion, raging among young ladies, to be peering through eyeglasses, not in any respect necessary, and universally known to be for the purpose of giving the wearer an imagined improved personal appearance, twenty years hence there will be some singular anomalies in female vision. There will be elderly ladies whose two eyes will not agree in focal axis. One eye will be long and the other short-sighted, the effect of squinting through a glass with one while the other is closed.

Possibly the difficulty may then be met by wearing glasses whose convexities are segments of spheres of different diameters.

It is for the future comfort, as it is for the preservation of their good looks, for ladies to use their eyes as they were intended to be used, together and not one at a time.

This unaccountable propensity for glasses, and to use them on the most frivolous pretences, has been a direct cause of thousands of defective eyes.

Since the introduction of gaslight in dwellings, various inroads upon vision have been recognized that were unknown in the days of candles and lamps. Oculists find their support in cities, particularly where gas and glasses are in the ascendant, and not in the country, where primitive customs still prevail in respect to lighting apartments.

Reading or sewing by gaslight, which is too brilliant, requiring protecting apparatus for shading the eyes, is far more trying to them than the old-fashioned lights. The oxygen of the room is rapidly consumed by gas-burners, leaving a sort of smarting sensation and a more rapid evaporation of the tears. We rub them, unconsciously, which promotes a more copious lachrymal secretion, which is temporary relief.

Drummond lights, gas reflectors, or a profusion of mirrors, gilded frames, and other reflecting surfaces in gas-lighted apartments in common family occupancy, are extremely injurious. Such continued stimulus of concentrated luminous rays produces internal inflammations of delicate tissues, and engorgements of vessels, which culminate in defective vision. All these sources of derangement give importance to ophthalmic surgery.

Gazing at grates of red-hot coal, as many do in their moments of mental abstraction, examining pictures by a vivid light through a strong magnifier, sitting in rooms habitually draped and carpeted in bright scarlet colors, and reading in rapidly moving cars,—are all of them destructive to distinct vision, and should be carefully avoided.

Furniture upholstered with dark colors, and carpets and curtains in which those shades predominate, are of far more importance where there are children, than has been suspected. Weak eyes, and even severe maladies, are sometimes due to such unsuspected sources.

# CHAPTER XX.

## Their Teeth.

Hereditarily Good or Defective—Hot Food—Smoking—Use no Dentifrices of
a Doubtful Character—Those most Useful—Quack Dentists—Employ
Men of Science—Cause of Caries—A National Characteristic, etc., etc.

Many allusions and cautions have already been given in
regard to the preservation of the teeth. But some more extended observations may be of service to those who have not
given much attention to the subject.

A hereditary tendency to an early loss of those important
organs is quite common; and when it does exist, no course of
medication is of much value in arresting the progress of decay.
It is possible to retard their early destruction by precautionary
measures, but they cannot be saved in their original appearance
of strength and beauty of structure.

It is within the course of general observation that defective
teeth are more common in towns than the country. Different
systems of cookery, condiments, and seasonings, together with
the custom of taking coffee, tea, chocolate, and almost every
dish that comes upon a table very warm, if not really hot, are
just so many agencies acting directly upon the enamel, till
openings are made through it to the bony structure of the body
of the teeth.

Hot food, ravenous haste in eating under the plea of urgency
of business, and hot drinks habitually, are unfavorable to the
health of human teeth. If by their organization they resist
such influences · through a long life, as they do with some

persons, it only proves their powers of resistance are stronger in some than in others.

The tendency of hot food and table-drinks is to disease the gums rather than the teeth themselves, in those in whom they remain sound, but seem to rise slowly out of their sockets in elderly persons. They are also thrown off by the absorption of the bony cell in which the fangs are imbedded.

Each root has a minute orifice at its extreme point, through which enters a nerve, an artery, and by their side, a vein to bring back the blood sent in by the artery.

In the body of the tooth is a cavity in which the nerve expands in a delicate plexus, which is the seat of exquisite pain when invaded, in consequence of the crumbling away of the walls which protected it.

### DESTRUCTION OF THE ENAMEL.

No branch of the dental profession has exercised the mechanical ingenuity of operators more than devising methods for preventing that calamity. If consulted early, when the first approaches of caries are discoverable, the arrest of the disease should be tried. Gold-fillings have the approval of the most experienced dentists. Various substitutes have been prepared and had a trial, but gold holds its reputation for superiority, and is not likely to be superseded. Why it is better than amalgams, artificial bone-paste, tin, or any other metallic filling, must be sought for in the publications and teaching of dental associations and colleges.

Through those decayed openings, sugar, cold water, etc., cause excruciating misery. When the pulp has been once invaded, it is rarely ever afterwards so secured as not to give frequent intimations of its sensitiveness.

Cracking nuts with the teeth, by no means an uncommon

vulgarity, is an abuse that may derange their connection in the sockets. There are so many ways of impairing the utility of teeth, it is quite hopeless to attempt enumerating them.

Domesticated animals fed on warm slops at distilleries, on kitchen refuse warmed and thickened with meal, with an expectation of increasing the quantity of milk—quite common with families keeping a single cow—do their pet incalculable injury. Cracks and exfoliations of the enamel follow such feeding. The perpendicular chisels that stand up in their teeth, of pure hard enamel, crumble and become black. Cold food is safest for them.

## TOBACCO.

An unfortunate opinion prevails extensively, that chewing tobacco preserves teeth. It is a popular error that has made many a toothless jaw. Grit, inseparable from the weed in curing, gradually wears down the teeth by the constant grinding motion, so that some men are met with in whom the tops of the teeth are nearly level with their tumid gums.

Women, happily, are not prone to that abominable, filthy vice of chewing tobacco; but they occasionally indulge in some of the Southern portions of this country in habits as reprehensible and obnoxious. They rub their gums with pulverized tobacco till it produces an agreeable sensation something like inhaling a few inspirations of chloroform. It is applied artistically with a brush, quite frequently when the habit has been established. Ladies smoke in Cuba. Some dilapidated females practise the same disgusting custom with us, but that circumstance does not lessen the objections that might be arrayed against it.

Tobacco-chewing is exceedingly offensive to those who do not use it. Chewers are nuisances everywhere, and especially

in public conveyances and private houses.  Floors saturated
with saliva, charged with tobacco and spittoons—an American
contrivance for protecting carpets—are sources of disease.
Breathing air in apartments where evaporation of such narcotic
filth is going on, must be exceedingly prejudicial, and, if carefully investigated, no.doubt, would be found to be the immediate cause of strange effects upon individuals of delicate
organizations.

There are women who virtually unsex themselves by copying the habits of men of low degree, in the use of tobacco.
Taking snuff is one of their bad imitations.  It is considered
unfortunate not to be handsome ; and old age with its wrinkles,
is dreaded by all women.  But that a homely one should take
to snuff is perfectly surprising, as she thus forfeits all hope of
being an object of interest, even to a Hottentot.

### DENTIFRICES.

Place no confidence in dentifrices, the composition of which
is a secret.  In this age of science it is a privilege to know
precisely what we use as food, in food, and for medicine.  It is
prudent to know too, what we are using for our teeth.

When preparations for cleaning teeth are secret compositions, beware of them.  Probably they contain an acid that
would gnaw into the enamel, or discolor the teeth beyond the
possibility of restoration to their primitive whiteness.

Teeth should not be brushed either with pulverized charcoal or pumice-stone, yet both are largely sold for that purpose.
They insensibly wear away the enamel.  To file off dark spots
would be precisely analogous, only the latter would be quickly
accomplished, while the other would be a gradual process.
Thus fluids would reach the bony structure, followed by dis-
coloration, decay, and tooth-ache.

Detergent soaps are allowable, being soft and free from grit. With a soft, flexible brush, soap, with cold water, removes adhering particles of food, and prevents the accumulation of tartar about the margin of the gums.

Immediately on leaving the table, it should be an established habit to cleanse one's teeth in that manner. Spasmodic attentions are to no purpose. Doing it when the thought occurs that they have been neglected, does but little good. It is by daily care that they are preserved.

When omitting to brush the teeth, even for a few days, with some persons, parasites actually burrow about their necks, and build up strong domiciles of calcareous matter, which destroys the periostic connections between them and their alveolar sockets.

Tartar, as it is called, a product almost as hard as coral, inhabited, too, by minute beings, which, under a microscope, exhibit active habits, should not be permitted to establish colonies in the mouth.

### DENTISTS.

When caries appears, consult a dentist, and be careful to employ no second-rate one, because his charges are low. There are dental institutions and colleges where the whole art and science of dentistry is taught thoroughly. Allow no cheap operator to prescribe or place an instrument on the teeth. Neither permit amalgams of mercury, copper, lead, or indeed any filling, to be pressed into a hollow tooth, which has not the approval of the magnates of the profession.

There are quack dentists, who rank next to quack doctors. Risk neither health, teeth, nor purse with either. Strange as it may appear, there are thousands who place themselves at the mercy of medical and dental pretenders, who would trust neither with their wallets.

A gentleman of New York, a little time since, consulted a medical gentleman on account of sore, inflamed gums, tongue and fauces. They had resisted a variety of medications till the gravity of the case alarmed the patient, and almost destroyed his confidence in the science of medicine.

After examination, shocked at the raw, inflamed appearance of the patient's mouth, looking as though burned, the doctor inquired whether he had any defective teeth. On reflection, he remembered that he had never discovered but a single decay in one of the back teeth, a long while before, which was promptly filled, so that his teeth might be considered perfectly sound.

The physician at once suspected the cause of such extensive disease of the mucous membrane had its origin there, and advised the immediate removal of the filling, and refilling with gold. His recovery and perfect restoration was immediate, showing there was a metallic poison in the first filling that had caused him so much inconvenience and suffering.

### NATIONALITIES IN REGARD TO TEETH.

Dr. John Allen, a learned, skilful dentist of New York, has collected an immense amount of valuable information respecting the history of teeth.

The body of a man, says Dr. Allen, with all its different parts, is composed of only a few simple materials combined in certain proportions to give strength and utility to the whole structure. Those materials are component parts of the food, and, although nutrient substances used by the inhabitants of different parts of the globe, appear quite similar, yet the food provided for them in various countries possesses the same general constituents everywhere essential to human organism.

Albanians of lesser Asia live principally on milk, cheese, eggs, olives, and vegetables. Sometimes they bake bread, but often eat their corn or maize boiled. Hippocrates says they were very strong in his day. Muscular, with oval faces, ruddy cheeks, and an animated eye. They had well-proportioned mouths and fine teeth.

In Central America, north of the Equator, the Mandingos have a barbarous custom of filing their front teeth to a point. The same extraordinary operation is extensively practised among tribes in various parts of pagan Africa.

In Eastern Africa, particularly, the Abyssinians have beautiful teeth, white and regular. Nubians, and residents of countries between Abyssinia and Egypt, distinguished for personal symmetry, having a dark-brown complexion, also are remarkable for their sound, white, strong teeth.

In Western Africa, and also in parts of Southern Africa, including Congo, the negroes are well made, extremely black, but noted for their superior teeth.

A people of that same vast continent, known as Khonds, of a dark color, straight and well-proportioned, are also remarkable for teeth of a pearly whiteness.

Turkish tribes of Kiptschak, the Tartars of Kasan, and all through that extensive region in the occupancy of bold, warlike, indomitably active men, are quite as celebrated for fine teeth as for their martial energy and determination of character.

Travellers represent the inhabitants of Eastern Arabia as being above the average stature of Europeans of a temperate zone. They are robust and active. With oval faces, copper-colored broad foreheads, black, bushy eyebrows, dark eyes, quick and restless, their sound white teeth are a remarkable national characteristic.

Arabs generally have sound teeth, even in the jaws, and rarely irregular. Unless addicted to chewing betel, they wear through a long life unimpaired.

Between China and Hindostan, the Siamese blacken their teeth, and also redden the inside of their mouths with a masticatory of lime, caoutchouc, and betel, which (says Dr. Allen), gives them a disgusting appearance.

Betel-chewing is practised extensively among the fellahs of Upper Egypt. Their lips and gums look as though they had been recently burned with a hot iron. Their teeth wear down level with the gums in a few years.

Tahitans have splendidly developed teeth, but they have an abominable custom of interfering with them, under certain circumstances—such as extracting some of them. If unmolested, they endure white and perfect to extreme old age.

New Zealanders do not exceed the common stature of , Europeans, and, in general, are not so well made about the limbs. Their color is of a different cast, varying from a pretty deep black to yellowish, with tolerably regular features. Their faces are round, with full lips, large eyes, black hair, straight and strong. Like most barbarians, their teeth are broad, fully developed, and white.

Capt. Fitzroy says of the New Zealanders, they are like those of the Tangians in regard to their dental apparatus. In old age they are either all worn down, or present an anomalous appearance.

Those natives residing near hot sulphurous springs or sulphur waters, on the borders of the lake of Roturna, have enamel on their front teeth yellow, although that does not impair their soundness.

To the eastward of the Society Islands, in the South Pacific, are the Gambier Islands. They are inhabited by a people fairer than the Sandwich Islanders. The average height of the men is about that of Englishmen, but they are not very robust. In their muscles there is a flabbiness, and in old men a laxity of integuments: their skin hangs in folds on different parts of the body. They have Asiatic countenances, with extremely white teeth; but they are represented to fall out at an early period.

In Easter Island, the most remote from the continent of all inhabited islands on the earth, there are finely developed inhabitants, with excellent features. The women are particularly handsome. Such beautiful teeth are nowhere else to be found. In the San War group of islands, all the natives have superb teeth. The Tarawan Islands, abounding in cocoanuts, fish, guava, banian trees, and sugar cane, the people have sound, white teeth.

The Great Vita, one of a group of islands between the fifth and nineteenth degrees south latitude, the inhabitants are celebrated for their sound teeth. So are the Feejeans. In fact, it has been the remark of voyagers generally, that the teeth of those distant islanders are always sound, white, and nearly as perfect as such organs can be, and remain so to extreme old age.

Vanikora, another cluster of islands, is inhabited by a black race, who cultivate taro, iguanas, and kava. Although small in size, they approach the negro in general physical appearance and organization, with countenances singularly resembling the ourang-outang,—their eyes being large, deeply set, and very much like those of the genuine negro of the tropics. Their lips are large and their hair crisp. An inveterate use of betel destroys their teeth early, which would last as long as those of the islanders of whom we have been speaking, were it not for the vice of chewing that abominable product of the vegetable kingdom which destroys them.

The natives of Australia differ from every other race of men in features, complexion, habits, and language. They have black hair, a cinnamon colored skin, and a dilated nose, with high cheek-bones—often an elongated upper jaw, with large sound teeth, very rarely defective in any respect.

Throughout South America, and everywhere on the Pacific, all tribes which have been met with from the earliest period of Spanish exploration, are distinguished for sound teeth. Exhumed skulls exhibiting a condition of the ancestors of all the tribes for more than a thousand years before any of them were known to European navigators, show what perfect teeth they had when living. Even after a lapse of ten centuries, they are still white, sound, and powerfully strong.

All aborigines of North America had sound, white teeth. Natives of Eastern Patagonia, according to Dr. Allen's memoranda, are a tall, extremely stout race of men. They are of a rich brown, rather of a reddish tint, with broad heads, rather flat on the top, a large mouth, thick lips, and prodigiously strong teeth. In one of the islands in the Magellanic archipelago, where the men are not more than five feet tall on an average, they are quite as remarkable as any race yet discovered, for white, sound, well-proportioned teeth.

In such estimation are sound teeth among some South American Indians, that they actually wear collars ornamented with them. Those strange appendages of humanity are called Botacudos. The Chaymas, another wild race, very analagous in physical appearance and similar in the practice of rites and ceremonies, in the estimation of Humboldt, leading a very simple life, have fine white teeth.

## CIVILIZATION IN REFERENCE TO TEETH.

Civilization has been destructive to teeth. A few, out of many, resist those influences which bring on premature decay; but a majority of the population throughout the United States have either lost some or the whole in both jaws. Where are we to look for a cause of such universally defective teeth ?

Dr. Allen is emphatic in denouncing the flour of which our bread is usually made, as the reason why teeth fall into decay. If flour were not bolted, but baked as it comes from between the stones in grinding, elements essential to the growth and reparation of the teeth would be disposed of in the system for their benefit. But the phosphate of lime, existing alone in the bran, is completely taken out in the process of bolting, leaving nothing for the teeth. That is fed to horses, swine, and cattle, whose teeth get the benefit of it, while we seek assistance of dentists, which would not be necessary, had we subsisted on food that had not been deprived of elements introduced in it to keep the teeth in sound working order.

Dr. Allen, closing his valuable researches on the anatomy and general economy of the teeth, expresses himself as follows:—

" According to our national statistics—1860—there were in the United States, 13,868 milling establishments for the manufacture of flour and meal, requiring 27,626 men, at an annual cost for labor of $8,721,391. Thus you see the number of men, mills, bolting-cloths, and dollars, that are employed in this great improvement devised by man for changing the proportions of one of the most important constituents in the country.

" The result of ignoring this mineral element from the staff of life is, undoubtedly, to a great extent, one of the most prominent causes of this national calamity (poor teeth), that sweeps from the population 20,000,000 of teeth every year.

"The potter cannot make the bowl without the clay, neither can good teeth be formed without a due proportion of lime, which is abundantly provided for our use upon the outer portion of the grain; and in rejecting that portion of the cereals, we virtually refuse to use the requisite materials of which the teeth are formed. We also deprive ourselves of a due proportion of atmospheric constituents, especially in our crowded cities. And also of the requisite amount of exercise to promote vigorous health and good constitutions. If we would be instrumental in doing more good in our profession, let us do all in our power to diffuse these important truths among the people."

In order to form good teeth, the proper materials must be used to make them; otherwise they will be defective in their structure, and liable to early decay.

The materials of which good teeth are formed are as follows :—

Phosphate of lime, with traces of fluoride of calcium ....... 67.72
Carbonate of lime.......................................... 3.36
Soluble salts.......... ................................... 0.83
Cartilage................................................. 27.61
Fat....................... ............................... 0.40

The enamel or external covering of the teeth has a still larger proportion of the phosphate and carbonate of lime. These different constituents are furnished us in the food designed for our use. Other constituents are also thus provided, of which the soft tissues are formed. Although there are traces of the mineral element in other articles of diet, yet the largest supplies are found in the cereals, in the following proportions :—

In 500 lbs. of whole grain (wheat) there is

    Muscle material...................................... 78 lbs.

    Bone and teeth material............................. 85 "

    Fat principle........................................ 12 "

500 lbs. of fine flour contain muscle material.............. 65 "

    Bone and teeth material............................. 30 "

    Fat principle........................................ 10 "

The Creator has not only provided the proper materials for building up the human system with all its parts, but he has also given us a fixed standard of proportions for each material to be used, which we should recognize as correct; but instead of doing so, we change the proportions of the mineral element (which is deposited in the outer portion of the grain) by bolting out nearly two-thirds of it from every barrel of flour, and discarding it from the staff of life, simply because it is the fashion to have our bread made of the finest flour, that it may be white instead of dark.

Now, it is estimated that a healthy child consumes half a barrel of flour in a year; and if this be fine, white flour, the child is denied twenty pounds a year of that portion of the grain which contains the proper materials for bones and teeth. This deficiency of the mineral element in the food causes the teeth to be comparatively soft and chalky in their structure; · and the result is, in this country, where fine flour is principally used for bread, there is not one in twenty without more or less decayed teeth before they have passed the morning of life. On the other hand, those nations who do not change the proportions of the mineral constituents in their food, do not lose their teeth from decay. This fact is well established by various writers upon the physical history of man, in different parts of the world, and is a recognized principle of physiology; and yet, as a nation, we are regardless of the consequences, and sacrifice

many millions of teeth annually. This national calamity can
be prevented to a great extent by simply popularizing a change
of fashion. Let the bread of this nation be made from unbolted flour. Let us cease to change the fixed standard of proportions in the constituents from which the teeth are made, and
then we may expect these organs to be well formed, and to
last as long as the other parts of the system. If this love of
fashion has too strong a hold upon the public mind to do this,
let parents, who regard the welfare of their children, ponder
well this subject, and decide which is best for their little ones—
fine flour or fine teeth.

The essence of all arguments advanced to prove that our
teeth decay prematurely, in consequence of the ill-treatment
they receive, has been printed and promulgated from so many
reliable sources, that it is lamentable no heed is given to
such important information.

Phosphate of lime, which is essential to the good condition
of teeth, is carefully sifted out of flour that bakers may have
white bread to sell. The bran contains it. That being considered of no real value, though a little better than nothing, is
given to swine, cows, and horses. Therefore there is fed out
to domestic animals the most important element in grain,
which, if used in human food, would insure better teeth and a
higher development of many silly brains.

Because this important fact is of immense consequence to
remember, that parents may pursue a course that might secure
sound teeth for their children, the statement, like some other
physiological lessons, has been often repeated in the foregoing
pages, at the risk of being considered unnecessarily tauto-
logical.

# CHAPTER XXI.

## Their Hair.

Women have fewer vices than men, but they have stronger prejudices. Whoever or whatever is liked they love; and whenever they hate, it is with the spite of a demon.

The opinions of women in regard to propriety and personal appearance allow of no interference; and in doing that which is actually detrimental to themselves, if satisfied it is the custom of a majority of the sex, they cannot be easily persuaded to change their sentiments. Reasoning is of no use with those who cannot be moved by arguments when they run counter to their wishes.

Women bear misfortune with heroism, but ridicule cannot be endured. Hunger, thirst, and innumerable privations are borne with becoming fortitude; but when they are objects of jest, in the way of derision, if no other way of escape presents, suicide is boldly perpetrated.

Nothing quite so completely engrosses their thoughts as dress. It is an idol of their adoration, and, therefore, an ever-present subject of contemplation. A woman unfashionably clothed had better be in a tomb, if she has aspirations for position. They also worship jewelry, especially in the form of rings, bracelets; and, above all, diamonds take such hold of them, that they are fashionably considered anchors which will hold a ship at her moorings through all the storms that threaten the stability of social life.

One of the first thoughts of a woman, whether a queen or a chambermaid of a second-rate hotel, is to have her hair tastefully dressed. Were the house on fire, or an enemy sacking the city, a true woman would flee with reluctance from impending ruin, if her coiffure were unfinished.

A woman's hair is an ornament which serves her longer than the flushes of health, and it would remain beautiful, thick, strong, and ornamental quite into advanced age, were it not badly treated. Because they are perpetually doing something to injure it, it is spoiled. As in the practice of other violations of sanitary laws, some individuals have such a fountain of vitality as to resist influences which destroy others; so in respect to the human hair. Some ladies are remarkable for its profusion and fine color late in life, while most of the sisterhood contrive to thin it out and destroy it, unknowingly, of course.

A woman's hair is an ornament, independently of an important service it performs in her vital economy.

## How Injured.

Because they are always endeavoring to improve its appearance by unremitting attentions, they are exceedingly apt to deprive themselves of the full development of a thickly-set head of hair by too much manipulation.

Some of the self-imposed cares which contemplate an improvement of their personal appearance, medicated washes, pomatums, etc., to their hair, do it an injury. Such violent discipline as it is subjected to with combs, not only breaks individual hairs, but inflammations are induced in the scalp which impair the office of the bulbs by raking the cuticular surface too severely.

Females so circumstanced by their low state of civilization as to rarely dress their tangled locks, have an immense growth of it. Squaws, particularly, who are habitually bare-headed in all conditions of weather, not only have a profusion of hair, but it is strong, long, and so well set, that even combing, a process only occasionally undertaken, neither loosens nor breaks it. Exposure, therefore, to the open air is exceedingly conducive to a healthy condition of that natural covering of the head, which performs an office in relation to the brain of which physiologists have as yet a very imperfect knowledge.

Not satisfied with giving a parallelism to hairs in combing, when masses are twisted into cords or closely braided, the strain given at the roots not only injures the cell from whence each hair springs, but the hair itself is maimed, and, its connection so disturbed, it becomes brittle, breaks easily, or falls out entirely.

This explains how the comb becomes laden with hair at each repetition of combing. Ladies are alarmed at it, and puzzle themselves for a reason of such a phenomenon. But nature would rarely be at fault, if its processes were not grossly interfered with by ruthless hands.

Women would become bald like men, were their bonnets as badly contrived as hats for excluding air. Being light, generally of open work, which gives a free ventilation, perspiration escapes, and an increased temperature of retained air is prevented. Then, again, they seldom cover the head, even with their light feathery gear, more than a few hours at a time in the course of twenty-four hours. The materials of which their bonnets are fabricated are of a texture far more favorable for the protection of hair, or rather non-interference with it,

than felt or stiffened glazed pasteboard, made impervious by a
coating of gum-shellac in all kinds of modern hats.

## HATS.

A few manufacturers, having become enlightened in regard
to the importance of having the same temperature within the
hat as outside, have small orifices made in the top or sides,
which no way mar its beauty.   Ventilation, secured by small
apertures, is philosophical; and had hats been so constructed
from boyhood, they would probably have saved many from
baldness whose heads have not a hair on the top.

## DESQUAMATIONS.

Desquamations of the scarfskin, in a mealy, white sort of
powder, under the common name of dandruff, is wholly due to a
protracted chronic inflammation of the scalp.   Successive crops
are thrown off, and they continue to be, just as long as the hair
is kept too much on the strain, by being pinioned with side-
combs and firmly-fixed pins.

 Whenever the slow state of inflammation continues
for a considerable time, patches of hair come out, leaving
bare, bald spots which are rarely ever reclothed with another
crop.   There is a vital tenacity in the bulbs which holds out so
that thin solitary hairs, short and sickly, give a hope of a
restoration, but they possess but little strength, and seldom
have much color or vigor.

Cases are cited, when, after partial baldness, new and vigorous hair shoots forth; but that depends more on the constitutional vigor of the individual than on drugs, pomatums, or other
miscalled hair-restoratives.

When hair does reappear, it is certain the cells which, in their aggregation, constitute a bulb, are intact. If they begin to secrete good hair, of a quality which was raised in youth, it must be a gratification, and the secret of it is the vital energy of the system.

Such bald places as have been described peculiar to women who bestow the most care upon their hair,—a reason for it is theoretically imagined to be the growing propensity of invisible parasites. But it is quite doubtful whether such mites are operating as extensively as supposed. In fact, whether any such destructive invisibles infest hair that is so often combed, brushed, and otherwise variously treated, is questionable.

## A BALD WOMAN.

A perfectly bald woman is extremely rare; still, there are a few. Those partially so are common. Wigs are so ingeniously fabricated, that it would be difficult to determine which has succeeded best, Nature or Art. A system of hair-dressing, commencing with the day, leaves it roped, cabled, and pinioned, as though each mass were a prisoner.

With such a condition of the head, an arrest of depilation could hardly be expected. The first step towards improving the secretion of hair, is to abandon severe tension, and the next measure should be to dispense with caps night or day.

A coarse net, merely sufficient to keep the hair from falling into disorder, is the only covering that should be worn. No tonic application will compare with pure cold water, next to air, which holds the first place. Its value is demonstrated in the immense development of hair on the heads of those who wear neither hats, caps, nor bonnets.

Some ladies are deluded with a theory that hair is kept soft, pliable, and glossy, by being covered with oiled silk. With that expectation, those more than usually solicitous for its pre-

servation, on the appearance of deterioration, fly to that pernicious course, and thus actually hasten a catastrophe they are most
anxious to avoid.

When a luxuriant growth of hair floats about so much unheeded by young misses as to be troublesome, the extent of
confinement to which it should be subjected is the use of a net.
Exhalations are not then impeded. If not as freely evaporated
from the cranial surface as from the neck, face, and hands, of
the roots of which such frequent mention has been made, will
surely take on a morbid action.

### EXTRAORDINARY GROWTH.

Very tall, slender, fragile young ladies, who develop prematurely,—that is, present all the physical signs of perfect
womanhood from thirteen to fifteen, are generally distinguished for a profusion of long, soft hair. It is related that one of
those delicate, and certainly too quickly-made women, who
leaped, as it were, from childhood into the full proportions of a
woman, without possessing a corresponding mental development,
had such an unnatural growth of hair as to cause her death.
It grew several inches in twenty-four hours, and consequently
exhausted the vitality of her system in an unprecedented manner. Such examples are rare, but occur frequently enough to
become matters of physiological record.

When it is apparent the development is in excess, the
quantity and growth of the hair being wholly disproportioned
to the rest of the body, and, therefore, self-evidently diverting
nutrition from other channels, medical counsel is advisable.

Advice from old women, just because they are old, is not
prudent. Hundreds of them in large communities are plethoric
with receipts for human afflictions; but neither their opinions,

nor their ointments,—commonly farragos of incompatibles, chemically considered,—should be accepted. They may succeed in making water-gruel, spreading mustard plasters, or understand the way of preparing catnip-tea ; but if health is a boon, never trust to any for prescriptions for preserving it, who are not conversant with the law of life.

## SIGN OF A VULGARIAN.

There are a plenty of bold men who might have been clothed in their own hair instead of a barber's wig, had they conformed to the usages of cultivated society—leaving their hats in the entry before entering a drawing-room. It is one of the rudest and most common of vulgarities, and therefore deserving a severe reprehension, that in some of the Southern States a man's hat is a permanent fixture to his head. Whether they are removed at night is a question. Certainly, they wear them in the presence of ladies as tenaciously as orthodox Israelites do theirs in a synagogue. If anything smacks of extreme vulgarity, it is to see a person claiming to be a gentleman, sitting in a parlor in conversation with ladies without removing his hat.

## ANTIQUITY OF WIGS.

Revelations from the mummy pits of Egypt show that subjects of the Pharaohs of the male gender all wore wigs. They were extremely light and skilfully made of delicate materials, which permitted a free ventilation. At present, and indeed for many centuries past, since mummy-making was abandoned, Orientals have their heads closely shaven about every ten days. Even male infants pass through the same operation, and have it continued as long as they live. As mummies were shaven as far back in the history of Nilolic civilization as any authentic evidences can be found, it appears that in exchanging wigs for the tarboush or red felt skull-cap—barbering the caput was not omitted. It is a national custom in the East of extreme antiquity.

It is an opinion, founded on the supposition, that vermin have always been such a source of personal annoyance in Egypt, that the only way of escaping from them was to cut off the hair where one variety principally burrow. Barbers are very common in the cities of Egypt, plying their razors on the heads of customers by the wayside at all hours. They use no soap, but simply moisten the hair with water, then pare the cranium as smoothe as an eggshell.

Females in that same vermin-infested country cultivate long hair like other women. They are less exposed to camels, donkies, dogs, and goats, than men, and hence less liable to the tribulation to which the other sex are exposed from their intercourse with those animals.

### PREMATURE LOSS.

A premature disappearance of hair, like a premature loss of teeth, results from neglect, or, in other words, in consequence of not taking proper care of either. It is asserted in a popular theological work, that teeth were never intended to ache. But they do, and generally those who deplore their loss are very much to blame.

Hair being a secretion directly from arterial blood through the agency of a peculiar glandular apparatus, intimately associated with cells from whence it shoots forth, if any violence is inflicted on them, their function is interrupted, and if that violence continue, they die and are obliterated.

A beautiful network of vessels and nerves surrounds each

hair bulb. The vascularity is apparent under microscopic inspection. Therefore, the less we do in dressing the hair beyond keeping it orderly, the better. By frequently cropping, it is supposed to thicken the hairs at their base, and encourage a more vigorous growth. To some extent, that may be true. When short, air is more freely admitted to the scalp, and insensible, perspirable emanations escape without raising an unhealthy temperature when pent up in ordinary hats, silk caps, and fur head-dresses.

The whole secret of having luxuriant hair is to keep it sufficiently loose for a free access of air, and never resorting to oils, pomatums, or bear's-grease, however sweetly scented to disguise their origin in lard, cotton-seed oil, or goose-grease.

No preparation compares with pure cold water, for giving a gloss and vitality to a lady's hair. Nothing equals it, and being within the reach of all, they have the means of securing a precious boon without money or price.

### GRAY HAIR.

Another point in regard to hair relates to its color. Ladies become gray, occasionally, while they consider themselves younger than they really are. It is no evidence of old age to have white or gray hair as early as when just emerging from their teens. It is a hereditary affair in such cases, and shows itself through one or two generations. Nor can a defective secretion of coloring matter be restored by any art or application known to science.

It has been said the Chinese have a mode of meeting the difficulty by taking something into the stomach that supplies the blood with an element for restoring the hair to its original color. The chemists doubt it, and they know quite as much and far more of science, than people of the flowery central kingdom.

Hair-dyes are extensively manufactured to cover up those premature indications of age, about which some ladies are extremely sensitive, without reflecting upon the fact that it is an incidental circumstance, sometimes quite independent of longevity.

Moral objections are urged against the use or resort to hair-dyes, on the score of its being a deception. But ladies practise other deceptions quite as heinous, and if one is wrong, the other is equally reprehensible; although no public censor has yet had the courage to particularize what those deceptions may be.

If the color of an edifice does not suit, the proprietor gives it another to meet his views, without causing any unpleasant comments of those passing by as to his right to interfere with a natural process of decay that is going on, or the moral terpitude of covering up a color which he does not like with another more acceptable to his taste.

Ladies have the same inalienable right to color their fiery red, yellow, or gray hair to black, brown, or any other tint which makes it more conformable to their individual standard of beauty, without scruple or apology.

It is a duty to look as well as we can to other eyes. If we can appear younger than we are by a little beet-juice on the cheeks, or have the hair at fifty look as it did at eighteen, there is no more wickedness in doing so, than in wearing artificial teeth.

If it is an offence in the sight of heaven to color our hair, it must be an offence also to substitute new clothing for thread-bare garments. The moment waning humanity attempts to rejuvenate in external appearance, there are troops of excessively good people who denounce it with holy horror, as a profanation and an unpardonable offence against Christian propriety.

We are advocates for harmless improvements in our external appearance, even if it relates to the substitution of new clothes for old ones. As the hair first indicates the decrease of vital force, there is nothing criminal, or particularly offensive to the public sentiment in keeping up cheerful appearances to conceal the melancholy discovery that we are no longer young.

No one is so weak as to suppose that by staining the cheeks or coloring the hair, either will prolong their stay on earth, or prevent them going to that far-off country from whence no traveller returns. To remain at a stand-still point and be forever in the bloom of youth, with no indications of having passed a meridian, cannot be expected. But it is gratifying to some to conceal their infirmities, but not so easy as to cover up wrinkles or mount a wig.

Through the instrumentalities of art, ladies succeed admirably in covering up many evidences of having been in the land of the living considerably longer than they are willing to acknowledge.

Pharmaceutical preparations for external and internal administration, of no value whatever, are articles of commercial importance, because they are represented to do so much towards the rejuvenation of antiquated females. They cannot be convinced of the imposition, so strong is the desire to appear in perpetual vigor. An active trade in hair-dyes, under the title of *restorers, regenerators, invigorators,* etc., therefore, is mainly sustained by those of both sexes who fancy gray hair speaks too plainly of age.

Do hair-dyes interfere with the health of those who apply them?

## HAIR-DYES.

Occasionally frightful accounts of their poisonous effects make excellent sensational paragraphs, and aid the sale of some new preparation that is represented to contain no injurious properties.

Under the impression that the skin absorbs fluids, hair-dyes are occasionally denounced. It is very questionable whether any cuticular imbibation can take place. Experiments, carefully conducted to determine whether it is possible for the skin to absorb any kind of fluid in which the whole body had been immersed for hours, and varied in temperature, it could not be detected in any of the organs, or in the secretions or excretions; nor by weighing before and after, was there any loss more than might reasonably be expected by evaporation.

Therefore, hair-dyes are not, and cannot be absorbed. It is possible to irritate the scalp with an acrid preparation. If there are abraded surfaces, cuts, scratches, or open ulcerations, then it is quite probable there might be both a local and a constitutional disturbance. But if no such conditions exist, then it is idle to dwell on the effects of a hair-dye, even if it is made up of such materials.

That lead may be held in solution to a very small extent in water drawn from lead pipes, in which it has remained considerable time, is not doubted. Some persons are extremely susceptible to its influence in the minutest form, while others are in no way molested by it. Thus, palsies are often traced to that source, and it is quite possible those supposed to have been partially or wholly paralyzed by hair-dye, received the lead in the water they drank, and not by its external absorption.

Most of the dyes have the reputation of being made of lead and largely of sulphur.

Lead pipes are objectionable, but it would be expensive to introduce a substitute. Millionaires only could afford to tap street mains with silver or glass tubes of sufficient strength to resist pressure from without or from within. Municipalities, boards of health, and chemists are convinced, by thorough investigation, that lead in solution in lead water pipes is so minute in quantity, as not to endanger public health. As well might printing be abandoned, and books and papers written with a pen, as before the invention of type, because one compositor in one hundred thousand has benumbed fingers in consequence of lead in their composition.

There is a vital chemistry—a preservative force constantly operating for the protection of the body—separating, carrying away, or neutralizing poisonous properties taken into the stomach in aliments and water, which, if allowed to remain unchanged but a little time, would be productive of painful consequences. It is in that way that lead poison is disposed of before so much of it accumulates as to become unmanageable by that conservative vital force which is a watchful guardian over organic life.

Some persons are infinitely more susceptible to certain impressions than others. On the whole, we must be reconciled to the contingencies of modern civilization. It would be absurd to abandon thousands of conveniences because it is possible some of them might raise a pimple, and thus mar the beauty of a fine face.

By no means run a needless risk in an effort to improve personal appearance; but in the application of hair-dyes, no danger need be apprehended, if the skin is not broken, and the scalp is free from ulcerations.

Simply moistening the hair cannot in any way conduct the fluid into the system. Hairs are not tubes which may be filled at their outward extremity, like a bottle; nor are they hollow cylinders, through which a stream may be conveyed to the skull. Hair performs no such function. The fluid of which they are formed is taken directly from arterial blood, flowing from the base outwardly. No inverted action can take place. Neuralgic twinges, numbness, or giddiness, from the use of hair-dyes, are not produced by its absorption. If at all, it is by evaporation, and inhaling the vapor into the lungs, and thus conducting the poison to the circulation.

Hair-dyes contain sulphur. That, too, is denounced on the false supposition that it creeps insidiously through the hair, like a sand-gigger, into the system. Irritation is not absorption. Continued applications of solutions of lead or sulphur would unquestionably become irritants, and they indirectly affect the general health. A palsy of the muscles about the eyes, or sides of the face, or of the broad, flat occipito frontalis that covers the top of the head, from hair-dyes, must be extremely rare; and then, rather from sympathy, than a direct action of the dye to the extreme twigs of the first and second branches of the fifth pair of nerves, which are finely dispersed in the facial muscles.

Nerves of motion emanate from the vertebral column, while those of special sense have their origin in the brain. Threads of the superior, middle, and inferior facial nerves which control the muscles on the sides of the face, are wholly beyond and independent of twigs from quite another source distributed to the hair bulbs.

There is no valid physiological objection, therefore, nor pathological, to staining white hair black, brown, or yellow. If mixtures contain pulverized cantharides, or any powerful irritant, of course uncomfortable consequences will follow.

Some hair-dye manufacturers claim that they restore the

hair to its original color. That is a mistake, since it can only be accomplished by a natural process, the tint being carried to the bulbs by arterial action. Chemically changing the color is not a restoration, and, besides, it fades out in a few days, if neglected.

Vegetable dyes are always preferable to metallic. Turks, Persians, Egyptians, and other Orientals, who glory in their intensely black beards, have them stained by decoctions and inspissated juices of plants. Inmates of harems, too, avail themselves of simple products of the vegetable kingdom for their raven locks.

Sulphur is extensively used in American hair-dyes, but that need not excite alarm or apprehension. When applied, as it often is, to the whole surface of the body in the form of an unguent for cuticular affections, in baths or internally, no baneful effects follow. How much less direct, when simply applied to the hair.

Whoever gives to the patronizing-hair-dyeing public a purely vegetable coloring fluid, will reap a rich return.

# CHAPTER XXII.

## THEIR FEET.

HAVING adverted to the painful consequences of wearing garments that fit too closely about the chest, without the remotest expectation of gaining converts among those for whom these observations have been written, the consideration of another evil of serious moment to the every-day comfort of women, is a subject not to be overlooked by them.

It being admitted that Nature is superior to art, it is extraordinary that women of sense continue to torture themselves, with an apparent resolution to compel Nature to sanction their follies. Notwithstanding the most positive and undeniable proofs that have been given in public lectures, and in printed volumes, to explain intelligibly the injurious effects of tight-lacing, women are as obstinately opposed to any change in that respect, as they would be to a revolution that would abridge their freedom, or interfere with cherished opinions in regard to their moral duties and obligations.

Feet were designed to be used in walking, and it must be admitted that their anatomical structure admirably fits them for sustaining the weight of the body.

An architectural arrangement of seven irregularly-shapen bones of the instep, brought together in a manner to form two arches, unequalled in strength and adaptation for the purposes contemplated in their structure, needs only to be examined attentively to convince a sceptic that the evidence of design is too forcibly demonstrated in the mechanical adjustment of that part of the foot, to be questioned by a sane mind.

There was a necessity contemplated for giving the lower extremities peculiar strength. The feet are complicated machines, managed by a multitude of vessels, cords, nerves, ligaments and vóluntary muscles, and yet, with all their complexity, if not ill-treated, they rarely get out of order. They would outlast some of the apparently higher organs, and are always in readiness for use when properly treated.

### Dissatisfied with Nature.

Women are notorious for being dissatisfied with that part of their own organization. Some of the kindest-hearted, sympathetic ladies, are intolerably severe upon their own feet, which they torture without remorse, when it would distress them painfully to witness the struggles of a fly in a spider's web. They comment, without apology, on the feet of other women.

They are harder upon their own feet than on the doubtful reputation of a rival, and recklessly tamper with their pedal extremities to their own discomfiture.

A small foot is more prized by some women than a full purse. She is a bold female who prefers comfortable shoes, if they appear to be large, while millions court the applause of fools who pretend to idolize little feet.

Laws of proportion are studied by artists in living beings; one may have a large head, another long fingers, a plump hand, or a coarse and angular pair of shoulders. Some are dis-

tinguished for short limbs, others are stilted up on immensely
slender legs, hardly larger than the slender supports of a
flamingo.

As people vary in dimensions, weight, strength or gracefulness, so their feet vary, but they are always precisely of the
size they ought to be, to sustain the pressure from above.
Unfortunately, ladies, as a general observation, do not see things
in that light. An arbitrary ruling of the votaries of fashion
has decided that feet must be small to be elegant. This is the
reason why distorted feet are almost universal among women
who are removed, by fortunate circumstances, above the lower
stratum of society.

They patiently submit to severe grievances without complaining, but if their feet happen to be larger than the standard
of gentility requires, their lamentations, though not always
audible, are, nevertheless, nursed in secret through years of
hope and ambition to be remoulded.

From the vanity of some ladies, whose thoughts are more
concentrated on their feet than their education, it impresses
spectators with an idea that they think more of them than they
do of the culture of their minds.

### ABNORMAL CONDITIONS.

Corns, bunions, incurvated nails, callosities on the heels,
riding toes, distortions, chilblains, and many other troubles of a
less grievous character, are each and all of them the results of
wearing such shoes as most commonly do not fit them, in consequence of determining to wear those which are too small.
Tight shoes are the immediate agents in the production of all
those pedal woes.

If small feet have their worshippers in worthless admirers,

there are those who view with sorrow a deplorable progress of that phase in civilization which cripples women in order to make them satisfied with themselves.

Swellings, œdematous enlargements of the joints, especially of the great toe, or a doughy fulness of the ankles, increased by too tightly-laced boots, are all the results of voluntary abuse.

By mistake, boots and shoes may be too small, but where there is a determination to conform to a prescribed standard, the wearer is not to be bluffed off by pain, or the outcry of oppressed flesh and blood at the point of pressure.

Continued compression cannot be endured long without disarranging the anatomical relations of the bones. The foot is built up of twenty-six bones, no two being alike, of the same weight, size, or shape, besides one or two additional ones, not always constant, called *sesamoids*, resembling split peas.

### DEVELOPMENT OF EXTRA BONES.

An extra bone may be generated to meet certain contingencies. These sesamoids are two, three, or even four in number, depending on circumstances affecting the particular region where an extra bone may be developed.

Originally, only two exist, and these are at the base of the large toe, being props for lifting the long flexor tendon farther from the articulation, to increase its power. Bunions are an inflamed thickening of the periosteum and an enlargement of the ends of two bones making the great toe joint.

If there is pressure at that point, long-continued inflammation sets in. The irritation extends down from the skin to the periosteum,—the membrane immediately investing the bone, which thickens, becomes puffy and exceedingly sensitive when its vitality is roused.

Each tissue is thickened, and the exterior becomes red and painful. Unless all pressure is immediately removed and applications made to reduce the inflammation, matter forms and sometimes is copiously discharged.

If not opened with a lancet,—the pent-up matter being allowed to remain, the bone may become diseased, which greatly complicates the misfortune.

Ulceration leaves the joint a little enlarged, even when treated skilfully. No shoe ever fits precisely or easily, after the periosteum has once been roused to inflammation.

Topical applications are only temporary relief. It is preposterous to think of a radical cure without removing the cause.

Another tribulation connected with uncured bunions, is the spongy enlargement of the long metatarsal bone to which the great toe is attached. Once enlarged, it seldom ever falls back to its normal dimensions. As a natural consequence, a shoe worn over it reveals the distortion, giving that part of the foot an unsymmetrical appearance.

Persons are constantly met with one or both large toe-joints so much enlarged as to immensely distort the shoe. Every step is attended with torture.

When a bone becomes diseased, there is an exaltation of vitality of a peculiar character. In health it is of a low order, just sufficient to connect it with a living system, otherwise it would stand in the relation of a foreign body, not to be repaired when injured or governed by laws of the general economy.

The nerves in the bones are extremely attenuated, while the circulation of arterial blood is sent to the remotest section, carrying in solution materials for growth or repair. Yet even such slender threads, communicating as they do with nervous centres, when contused or invaded, immediately communicate the fact. An injured bone cannot be pacified easily. Medications for them consist principally in topical appliances for reducing inflammatory action.

The severest sufferings from bunions or corns are not permanently relieved by unguents, emollient lotions, or paring away hardened cuticle. The remedy which is a cure, is simply wearing shoes that do not press on the tender spot.

Wearing sandals for a month, which have no vamps, would allow nature to reestablish order where it has been disturbed by tight shoes.

### TEMPORARY RELIEF.

A sensible way of seeking temporay relief practised by laborers, is to cut a piece out of the shoes or boot, over the bunion. A hole thus made, affords immediate relief from agonizing pain.

No outlay for advertised specifics need be expended. Freedom to the oppressed part is all that is required.

When ladies reach their dressing-rooms from a promenade, distressed by their beautifully-fitting boots, their first act is to exchange them for soft slippers—the older the better. Otherwise they sit in their stockings, shoeless, till the anguish brought on by exercise in their tormentors has somewhat subsided.

### CORNS.

Corns speak in forcible language, which makes those who have them realize that no half-way measures are successful in their treatment.

They actually spring into existence to defend the spot where they appear, from impending injury; and as faithful sentinels, cry out at every movement which menaces the locality under their charge.

Barefooted people have no such afflictions, nor those who wear cast-off shoes a size or two larger than their feet. Corns never appear unless the toes are wedged too closely. Being *pinched* expresses the condition which develops those painful prominences.

If the vamp of a shoe is too low, the pressure interferes with a free circulation on the upper surface of the toes. Inflammation follows, the cuticle begins to thicken and rise above the ordinary level.

On its underside or base, a corn has a conical shape—the point, like a thorn, on the slightest pressure, irritates the inflamed periosteum below, and thus they act as messengers, announcing through the nerve filaments something wrong is transpiring, which is thus telegraphed to the brain; which, if a sensible one, will remove the tormenting pressure.

In these pedal miseries, voluntarily induced, a demand is made for a distinct profession to meet the contingency. Thus chiropodists are in the enjoyment of lucrative incomes. Corn-doctors have a thriving business in cities.

Corn-martyrs do not deserve much commiseration, because they might have permanent relief by simply discarding tight shoes.

Softening corns in tepid water, and afterwards paring them down, is only temporary relief, with a moral certainty of a speedy uprising again to the former elevation. The more prominent, the worse it seems to dig down into the flesh below.

Corn doctors are not infallible. They promise well, but their operations must be frequently repeated.

## CHILBLAINS.

Chilblains—those burning red patches which are excessively irritable on the heel, the sides of the feet, and occasionally on the sole—are produced quite as often by pressure and the non-escapement of perspirable emanations from the feet as from snow-water.

Glazed leather and India rubber shoes and boots prevent the evaporation of perspiration from the feet, and hence they become extremely tender and liable to chilblains. India rubber constricts the toes, by tightening the bones and deranging their original relations.

Such shoes should only be worn for a very short period—for walking through muddy streets—and removed on entering the house. Aside from the injury inflicted on the feet by wearing them as some do, indiscreetly, days in succession, if the perspiration is pent up and not allowed to escape, the general health has been found to be disturbed from that cause. Habitually worn, India rubbers distort the feet and leave them extremely tender.

Thin shoes, too thin and light to resist moisture from without, particularly when there is snow on the ground, invite chilblains. Ladies should wear shoes as thick and strong as those worn by men, if they are similarly exposed in the open air. Thick soles ought not to be forgotten. Ordinarily, they are not much thicker than paper, which explains a liability to those erysipelic attacks which commence suddenly and run a rapid course.

## ERYSIPELAS OF THE FEET.

Solutions of common salt, sulphate of zinc, decoctions of rose leaves or camomile flowers, are each and all of them soothing, and not unfrequently effectual in dispersing the malady, if applied seasonably.

## EFFECTS OF CONTINUED COMPRESSION.

There is scarcely any difficulty or derangement of the toes or feet that does not originate in violence from compression. Cotton-batting, stuffed between the stocking and the corn or bunion, so as to raise the shoe above the corn, is an admirable way of obtaining immediate relief, when so circumstanced that no other more permanent treatment can be had.

## HIGH HEELS.

Sprains, abrasions of the skin, etc., which are inconveniences, may frequently be traced to immensely high heels which ladies cannot dispense with, who make pretensions to fashionable equipment.

With these warning words, if they still persist in making themselves uncomfortable even to intense suffering, they must be given over as incorrigible and willing dupes to the arbitrary demands of fashion which imposes hardships upon them greater than they ought to bear.

It will be an amusing exhibition for a distant generation to have pictorially illustrated the phases of female fashions of this generation. The cut of garments, high heels, enormous hip and other paddings, pyramids, of artificial hair piled on in such profusion as to be entirely out of all proportion to the rest of the body, with other ridiculous contrivances that must embarrass their freedom of motion—could not fail of being contemplated then as now, with mirthful astonishment.

## A GRACEFUL STEP.

A steady, dignified step, is hardly possible on high heels. Mounted thus, the weight is thrown forward, the shoe becoming an inclined plane, which gives a peculiar stoop that took the name of the Grecian bend when first introduced. It is like standing on the roof of a building, cobbled up on high heels, being continually obliged to resist a tendency to pitch forward.

High heels bring immediate trouble to the toes, by wedging them into the extreme point of the shoe.

Absurdities in dress die out for a while and then revive again, as though humanity could not be satisfied without being slightly miserable. Just at this particular juncture, high heels are very high, with a base of support not much broader than a finger-nail. Shop window specimens exhibit the sacrifices women make to appear taller than Dame Nature ordained them. Fashion or death is the ruling spirit, and some have both.

Female pedestrians step out of their high-heeled boots as quickly as possible on a return from a promenade. Heels, even half an inch high, cannot be worn without bringing an extra strain upon some of the muscles of the leg, particularly on the long flexors of the foot.

Ridicule heelless shoes of Orientals as we may, they are philosophically right, and we are wrong. They are at ease with them, while our ladies are only comfortable when they are off.

How much rheumatism, neuralgia, and cramps are due to high heels, may be ascertained by the study of works on morbid anatomy.

## PARTIAL ADAPTATION TO CIRCUMSTANCES.

By persistence in wrong-doing, that is, voluntarily making one's self uncomfortable, the muscles of the foot and leg after awhile adjust themselves partially to the new condition, but always at the expense of a loss of tone, and of the full exercise of their normal power. Whenever liberated, they contract back to their former state, which is a permanent relief, if not again compelled to act unnaturally.

Men are no wiser than women in regard to the high-heel mania. Their boots are elevated quite too much at the heel; consequently, they are familiar with corns and bunions, enlarged toe-nails, unsymmetrical feet, and bulging out of the leather over those irregularities, created by forcing the foot forward into a narrow extremity of space.

We were all born with good pedal extremities, precisely adapted to the plane of the earth, and they would serve us admirably, free from excrescences, incurvated nails, riding toes, callosities, protruding joints, and other annoyances, to extreme old age, if they were never put into unyielding leather prisons, too small to receive them.

## ANTIQUE FOOT.

A small foot may be exceedingly beautiful in the estimation of those who have very large ones. If narrow, and the toes are in close contact, the foot is not a true type of the best form. Sculpture represents the toes spread, so that there is space between them, thus giving them a firmer hold and a broader base of support.

Camels are born with callosities over several joints on which they rest while being laden or unladen. Man has a thick, compact protecting cuticle in the sole of the foot, beautifully cushioned for protecting nerves, blood-vessels, and tendons under an archway of small bones, between the heel and base of the toes. The arch is kept in place by inelastic ligaments, running from point to point, so remarkable in the disposition made of them, as to show, beyond the cavil of ingenious doubters, that intelligence was exercised in their distribution, to give perfection to the foot. Without just that particular arrangement, the weight of the body would crush the structure into confusion, and utterly destroy the mechanism.

As we did not contrive our own bodies, we must admit of the existence of a Supreme Intelligence that did produce such marvellous mechanism.

Artists find the foot a profound study, simply looking to its exterior; while anatomists are rapt in wonder and admiration at revelations in its interior.

Both elementary anatomy and physiology should be taught in female schools and seminaries, that the pupils might have early insight into their own complicated organization. It would make them more careful of themselves, and lead to the observance of those laws, the violation of which, through ignorance, whether relating to their stomachs, their brains, their eyes, or their feet, embitters life, and destroys them before they have had as much of life as they would have had under a more perfect system of education.

A medical gentleman of Boston excited considerable derision some years ago, because his common-sense was superior to fashionable folly. One of his so-called foolish whims was, in having shoes for his children made exactly to conform to their outline, marked round on a piece of leather with a pencil. Their shoes had a comical appearance, to be sure, contrasted with modern manufacture, but an object of importance was attained, viz., good, sound feet.

Let those laugh who win.

# CHAPTER XXIII.

### THEIR PHYSICAL NECESSITIES.

Is life essentially prolonged or shortened by the quality of our food?

Many physicians would answer no, if they gave the subject much thought. Each and all entertain theories which naturally have an origin in deliberations on the phases of disease, and the influence of diet.

Most persons have a vague notion in regard to themselves, in reference to what may or may not be suitable for the stomach.

Even those of eminent physiological attainments are often influenced by whims, rather than by facts, in their theories of life. Evidence is extant of the highest import, incontestably proving that it is of very little importance, or rather of no consequence, what kind of food we subsist upon. Longevity depends on a peculiar vital endowment, transmitted from parents to children. Neither food nor climate perceptibly modifies the life period, aside from outbreaks of pestilence and epidemics.

A beggar in the street lives as long as one who satisfies every craving of his nature. Wise, considerate, and learned men who believe themselves masters of hygienic laws, cannot arrest the progress of what is denominated self-limited disease. Nor is it easier to arrest the tendency to long life, when the food has been wholesome, without violence.

There must be specific laws regulating the life period of all animals and plants. With the aid of science, it is possible to acquire a better knowledge of those laws of limitation. It is within the sphere of possibility to determine the precise day of death.

## FEMALE MEDICAL EXAMINERS.

This subject has been studied with earnest solicitude by life insurance managers, but, unfortunately, the inquiry has too frequently been confided by those institutions to medical donkies, instead of men of brain. It is very mortifying that medical examiners appointed for the express purpose of discovering the physical prospects of life in applicants for policies, are not often distinguished for ability, educational acquirements, or professional standing. A man of knowledge, fitting him to counsel executive officers in granting the benefits of life insurance, has not much chance of appointment, unless he is a relative of some controlling spirit of the institution. Were researches made into the organization of many companies, it would surprise the public to learn they are family affairs, largely owned and managed for the support of a president, secretary, cashier, and other officers, including fathers-in-law, brothers-in-law, cousins, sons, and nephews, and occasionally doctors, all held together by a tether of consanguinity.

Women should be the medical examiners of women for life-policies. Reasons might be given for this assertion, of importance to companies. A female medical examiner should be attached to the office permanently, even if she were not a relative to the ruling elder.

## DIMINUTION OF VITAL FORCE.

When we have reached a period at which there is a full development of our powers and faculties, the scale is turned, and a diminution of life-force is gradually perceptible. It is precisely so with animals, in whatever climate they are located. There is less activity in the circulation, a gradual relaxation of the tissues, and an increasing obtuseness in the nerves of special sense. A reluctance to engaging in pursuits that formerly were sources of pleasure, is another observable circumstance, indicating a culmination and downward tendency of the body and mind.

Though there may be a long lingering old age, the day of doom at last arrives. Rude winds rend a limb here and there, and by and by a gale in its fury levels the giant oak with the ground from whence it came. As it is with stately trees of the forest, so it is with monsters of the deep. A whale may roam in the depths of the ocean for centuries, able to withstand terrific assaults of formidable enemies, but the great heart that drove a column of blood one hundred feet at each pulsation, finally beats for the last time, at the end of a thousand years, —for aught we know to the contrary—in obedience to a law of limitation.

Though we understand many of the once-called mysteries of Nature, yet we cannot ward off a blow that will terminate existence, when most solicitous to live. Man, of all created beings, has a conscious knowledge of what must transpire in regard to the close of life, without being able to avert it.

### SANITARY PRECAUTIONS.

Moses gave the first code of sanitary regulations ever promulgated, which are substantially in force at the present day in most Christian countries. Wherever they are strictly observed in respect to animal food, the people enjoy the best health.

Were a catalogue given of the kinds of food on which

humanity should subsist, it would not be satisfactory, simply because articles that would be excluded as dangerous in one country, might be valued as very superior in another.

But man being omnivorous, he can be sustained on anything which yields nutrition to graminivorous or carnivorous animals. In Arctic regions, the demand of the stomach is for fat meats and animal oils. On approaching the Tropics, both the quality and quantity is constantly varying, the craving being for a mixture of vegetable with animal food—the appetite for the first rather predominating.

At the Equator, fruits, grains, nuts, seeds, and roots are the principal food of the inhabitants; but, according to travellers, a desire for animal aliment becomes so perfectly uncontrollable at times, as to lead to terribly revolting exhibitions of cannibalism.

Disgusting feastings on human flesh are almost certain to take place every few months in the gloomy interior of that part of Africa which is rarely penetrated by white men,—the home of gorillas,—if meats cannot be procured from other sources.

A demand for elements, nowhere else found but in animal food, partially explains those barbarous acts of feeding on a fellow-being, which characterize the rudest condition of human society.

There really is no positive standard, that is, a catalogue of articles which are proper, and exclusively so, for nourishing the body.

Were a butcher to sell horse meat in our cities, he would, unquestionably, be prosecuted for vending an unwholesome article, unfit and unsuitable for human food. A feeling of intense exasperation would probably agitate the community where such an outrage had been perpetrated. Yet, in Paris, horse-beef is a recognized market production, and well esteemed as nutritious and proper. There were eight markets in which it was extensively sold before the late revolution.

Let a prosecution be commenced almost anywhere in the United States, against some one who had the hardihood to sell horse-meat, and, ten chances to one, there would be an array of medical experts to testify it was an infamous transaction, destructive to individuals, as it would be to the public health.

Man is omnivorous, and, because he is so, amply qualified to range over the globe, regardless of circumstances which restrict most animals to particular localities in which their appropriate nourishment is provided.

What would become of the inhabitants of Lapland, deprived of fish and seal,—no vegetables to be had there?

Necessity compels those at the Arctic Circle to feed on that which will best keep up the current of their vitality. Under another condition of climate, millions subsist on rice. But the intellectual calibre of both fall infinitely below those in temperate zones, who are sustained on a mixed diet of flesh and vegetables.

Our jaws are studded with four distinct kinds of teeth,—viz., incisors, or cutters, in front; canine, called eye-teeth, for tearing and holding firmly; single and double molars, exclusively for grinding.

Carnivorous animals have no grinders, the graminivorous are without the canine, as they appear in dogs, lions, tigers, and the like. The motion of their jaws is up and down, cutting upon the principle of shears, with no sliding movement. Cattle, horses, camels, etc., grind their food into pulp before swallowing it.

Man both cuts, rends, and grinds. In short, he performs all the acts in preparing food for the stomach, which the animals referred to perform singly. Thus, anatomically, is a proof found of his omnivorous nature.

Passing from the further consideration of the omnivorous character of man, to qualify him for a general superintendence of the earth's surface, it may be fearlessly asserted that those who confine themselves exclusively to a vegetable diet, will never be distinguished for their intellectual powers.

A flourish of trumpets and tempestuous declamations before weak-minded audiences of converts to any *ism* which happens to be promulgated by adventurers for notoriety, occasionally secure a disciple who is captivated with the announcement that we were designed to subsist exclusively on vegetables.

Their physical and mental deterioration begins when they adopt the system. A temporary brilliancy, and vaunted clearness of perception is imagined to result from an abandonment of animal food for baked apples, boiled turnips, and roasted potatoes.

Rapsodies from a change of habits are symptoms of approaching lunacy.

Women require a mixed diet. They should take, without reserve, whatever belongs to the family regimen. This is not to be construed into an arbitrary system of dietetics, from which no deviations are allowable. Whatever is relished and digestible, is proper.

Meats have been human food in all ages, and they will continue to be served while humanity remains the same.

If men were originally monkeys, they probably subsisted as monkeys now do, on nuts and farinaceous products. When men confine themselves exclusively to vegetable food, they will dwindle down again to the level of their putative ancestors.

A mixed food of animal and vegetable is a law of necessity in temperate zones.

# CHAPTER XXIV.

## Minor Sources of Annoyance.

Pride—Mutilations without Destroying the Intellect—Ligation of Limbs by Elastics—Freckles—Epidermis—Moth Patches—Nostrums—Grass Food —Danger of Topical Applications—Red Noses—Astringent Lotions— Smelling Bottles—Stimulants—Appearing to Advantage.

A NATURAL instinct urges us to appear to the best advantage before others. That leads to placidity of deportment, propriety of conduct, and the practice of courtesies which are agreeable, if not essential, to a good understanding with those with whom we are associated.

It would be hardly short of insanity to seek opportunities for disgusting acquaintances by habits offensive to decency or the common usages of well-regulated society.

This inborn disposition impels us to efforts for improvement, and to conceal defects, real or imaginary, that might diminish our attractive qualities. External appearances have more influence with the majority of mankind than intellectual attainments or moral excellences of character.

When pride is in excess, it eventuates in ridiculous exhibitions that provoke comment and biting remarks. With a desire to improve personal appearance, the remedy is not unfrequently worse than the defect. Thus wigs, cheek-plumpers to puff out hollow mouth-walls, artificial eyebrows, gum-elastic bosoms, wooden calves to spindle-shanks, and some other devices for appearing developed to a commendable standard of excellence, cannot escape comment when the deception has been discovered.

## MUTILATIONS POSSIBLE.

There is a story illustrative of the pruning a living human body may pass through, without destroying life or apparently impairing the mind.

When Miss Jones became Mrs. Brown, the happy husband was nearly frightened out of his senses by the extraordinary metamorphoses through which she passed. He had gazed with pride on Mrs. Brown's fine proportions..

Knowing her to be a woman of discretion, whose forty years of singleness had afforded ample opportunity for qualifying the charming creature for superintending the genteel establishment to which she had been matrimonially introduced, Mr. Brown congratulated himself on the prospects of his domestic future.

Retiring, Mrs. Brown first removed a splendid head of hair. Next, on taking off a pair of gold-bowed spectacles, out came one eye. Laying both on a table, she then deliberately withdrew a double set of milk-white teeth. Progressing, a full panting bosom was unbuckled. Taking a position before a mirror, one side of her porcelain nose came off. Sitting down, a wooden leg was unscrewed, and then the left arm just below the elbow!

Such are among the mutilations possible, without in the slightest appreciable manner interfering with mental operations.

All artificial appendages which improve the corporeal proportions, while contributing to the comfort and sometimes to the necessities of the individual, are allowable and should be encouraged. It is high art to so improve and conceal defects which are unpleasant objects to others.

Dentistry has largely contributed to the restoration of impaired faces, and essentially benefited millions whose digestion was defective from the loss of teeth.

## What has been Neglected.

Elastics for keeping sleeves and stockings in place have escaped observation. It is time they received attention from physiological reformers who devote themselves to teaching the way of long life by the avoidance of popular abuses, self-imposed and, therefore, the more difficult to remove.

Those girders obstruct a return of blood from the extremities, through superficial veins, and therefore should be abandoned.

A reason why some ladies have very small, bony limbs, is because they have not blood enough circulating in them.

Elastics below the knee block the cutaneous veins; and those articles under the name of sleeve-bands worn on the arms, obstruct the currents in both arteries and veins, as they press them against the bone.

Garters do not produce much pressure on the arteries, as they are deep-seated and protected from compression by their favorable location.

Hose should be kept up by elastic straps, a few inches in length, extending from a button on the drawers to another at the top of the stocking on each limb. That simple contrivance completely relieves the vessels. If the circulation is unimpeded, the limbs will develop under appropriate exercise.

## Freckles.

Freckles are regarded as afflictions. Persons of a light, florid complexion, especially those having redish, or entirely red hair, more generally than others, are apt to be marred with dingy discolorations of the skin.

Dark hair, dark eyes, and dark complexions are usually exempt from such anti-beauty spots.

Freckles cluster under the lower eyelids, by the sides of the nose, back of the hands, on the upper part of the neck, or, indeed, wherever there is an habitual exposure to sunlight in a particularly warm season.

Washes, lotions, teas, etc., without number, are everywhere on sale, represented as efficient in the removal of such defects. But they are utterly useless, doing damage rather than relieving the skin from offensive dingy discolorations, freckles, or yellow irregular patches.

Exclusion from solar light is a precaution, in the brightest part of the day. A veil is unquestionably a partial defence against intense rays of a brilliant sun, which corrugate the skin where the coloring pigment under it is thin or scantily secreted.

The epidermis, or first skin, is both thickened and corrugated at intervals of a few lines, by exposure to the sun's rays in many persons. Those of a nervous, sanguine temperament, and of a light complexion, are most susceptible to freckling influences.

It is consolatory to believe in the theory that freckles are protecting shields to parts immediately under them, particularly when the attempt to remove them is unsuccessful.

That freckles prevent the passage of some of the prismatic rays from reaching something that ought not to be impinged upon by them, is rather an assumption, than susceptible of proof.

Whether pores, or twigs of cutaneous nerves, are protected from injuries that might ensue, were it not for thickened places in the outer tissue, requires more and closer observation than the subject has hitherto received.

Freckles are both mechanical and chemical barriers to properties in sunlight that would inflict an injury, if not intercepted. Such is the imagined origin of them with those who have more imagination than facts to build upon.

Possibly, extreme minute capillary vessels are protected in their labors by being covered by a thicker scale—for such is a freckle. Where there. is one, it is a darker, thicker spot than the space between any two of them.

## MOTH SPOTS.

Moth-patches, as they are called, being irregularly defined discolorations of a yellowish hue, commonly appear about the chin, the base of the ears, on the forehead, and, indeed, just where they are conspicuously in sight: oftenest on the faces of ladies of a lax habit. Nursing women, and those who pass much of their time in poorly ventilated apartments, are most predisposed to such unwelcome appearances.

No calculation can be made respecting their duration. Young mothers are sometimes suddenly surprised by those yellow markings. Ladies, too, in middle life, without any assignable cause, are also the occasional subjects of moth-spots.

Quacks and nostrum-venders hold out encouragement for their removal by applications of secret compositions. But there is no utility in their farragos.

## NEW APPLICATION OF STEAM.

A process has been successfully practised of late for the removal of those disagreeable discolorations, which is unobjectionable, and far better than dosing with internal medicine that can have no efficacy whatever.

The place is covered by a cup, from the bottom of which extends an elastic tube communicating with a vessel generating steam. The hot vapor is thus applied at a bearable temperature, to thoroughly soften the skin to a point that it may easily be rubbed off by the finger on removing the cup.

That parboiling process also softens the pigment, which also slides off from the cutis vera, or true skin.

A return ·of the yellowish coloring matter may gradually reappear. Application of the steam-vapor a few times, at the same time circulating freely in open air, rarely fails of accomplishing the object.

On peeling off the mothy skin, cover the denuded surface with gold-beater's membrane or thin court-plaster, in order to exclude the air for a few days.

## Avoidance of Causes.

Gross food, such as too frequent meat-eating,-pepper, vinegar, or irregularities in diet, are thought to contribute to moth-spot development.

Pimples, elevated purple aureolar discs, minute vegetations near the wings of the nose, clusters of black dots, and hard, gnarly moles on the face, cannot always be removed without excision.

There is absolute danger from the topical application of so many falsely-named medicinal remedies; the compositions, when known, being invariably condemned by physicians. It is never safe to tamper with drugs of any kind; neither pills, powders, nor fluids, however prominently recommended, without approval of a medical adviser.

In taking preparations, the ingredients of which are unknown to any but those interested in the profits, difficulties are produced, not unfrequently far more serious than the minor ailment for which they were given.

## A RED NOSE.

A red nose on a lady's face is an extreme mortification. Sometimes an intense shining redness remains a fixture for months in succession, unaffected either by external or internal medications. An engorgement of cutaneous blood-vessels on the cheeks or nose resists discutient applications far more obstinately than inflammations on other parts of the body.

Sometimes the tip of the nose is of a shining tumid redness. The vessels of the skin are in a permanent state of inflammation. Cooling lotions rather aggravate than ameliorate the tumefaction, which is opposed to the theory that excessive local heat can be reduced by cold applications.

Lead water also aggravates the condition; and worse still, if persisted in too long, results in a loss of sensibility at the tip, by producing a paralysis of the cutaneous nerves distributed there.

For a red nose, mild treatment is safer and more successful than harsh measures. If air and light are excluded, very easily accomplished by a covering properly fitted through the night, and much of the time through the day, considerable relief may be anticipated.

But the best method to pursue is, to apply soft, emollient applications, mildly warm. Fine Indian-meal, in the form of a poultice, mixed in alum-water, should be worn through the night. Follow up the practice without intermission, for weeks. It is best not to have the mixture very astringent at first. The astringency may be gradually increased by a solution of more alum.

The skin becomes gradually softened, the enlarged vessels diminish, and, as the inflammation subsides, the redness disappears.

When the poultice is removed on rising in the morning, favor the partially-parboiled surface with a soft piece of oil-cloth, pierced with orifices for seeing, breathing, and using the mouth.

Avoid smelling-bottles, pungent odors, snuff, and all other irritants of the nasal cavities, when a tendency to an engorgement about the wings or the nose itself exists.

A caution in regard to liquors may be unnecessary to ladies ; however, it is certain that any extra excitement which drives the blood rapidly would be an aggravation of inflamed patches on the face.

Red-nosed smokers must abandon their idol, if they have a desire to recover their once good appearance. A volatilization of the essential oil, or whatever property is diffused in the smoke from a cigar or pipe, seems to add fuel to the inflammation.

Women being less prone to the use of stimulants than men, and less exposed to various demoralizing influences from pro fane and vulgar associations, escape many ills which are incident to weak and thoughtless men.

Women are occasionally seen with red noses, and morbidly flushed cheeks, who are egregiously imposed upon in their haste for relief. They are duped into purchasing vaunted specific remedies that have not the least medicinal virtue.

If ladies have more credulity than men, happily, they have fewer sins.

Many precious lives are sacrificed on the altar of female vanity, in the earnest pursuit of phantoms.

Every woman exerts herself to appear to the best advantage. That prompts them to appear neat and tidy in their persons, and if they have blemishes, real or imaginary, they strive to remedy the defect as speedily as possible.

That is why they are patrons of all sorts of advertised nostrums which promise more than can be performed. When all women are dead, there will be no more sale for patent medicines.

# CHAPTER XXV.

## Their Peculiar Organization.

To point out all the anatomical differences of the sexes, is not contemplated. No subject would be more difficult to popularize, and, were it accomplished, there would still be problems unsatisfactorily · managed to meet the theoretical views of those who are always ready with objections, even when nature bears witness to the statements and deductions of medical philosophers.

From childhood to age, there is a marked difference between men and women in their physical structure and appearance. Moral qualities are laid aside in this examination.

There is a delicacy in the very bones of a female, that contrasts singularly with the strong, hard, rough skeleton of the male. But in some of the carpentry of the osseous system, it is obvious that intelligent reference is manifested in the variations recognized by anatomists to specific purposes which do not exist in the male.

With the same number of bones, arranged in the same order, and fulfilling the same offices, and moved by precisely similar muscles, influenced by nerves exactly like those in man, yet a woman is not a man. She is of mankind, and yet she is by herself.

Though of a finer texture, and operated upon by subtle influences, regulated by a law of periodicity past finding out, she is really no more complex than a strong, athletic barbarian in her physical economy.

A woman is not a perfect being by herself, neither is a man. The two constitute one, and that is a relation contemplated from the beginning by the Power that fashioned them. From birth up to a pubert age, some parts of their system have been so slowly developing, the physiologist is perplexed in his attempts to make plain an interesting chapter regarding the phenomena of development.

While the brain enlarges in volume, the limbs lengthen, the muscles increase in bulk and strength, essential organs in the economy of animal life remain quiescent for years. Perhaps it is better to say *apparently at rest*—performing no office for a long while. That apparent quiescence is undoubtedly a period of extraordinary changes, with reference to a revolution which changes the child to a woman.

Precisely so are the conditions of boys through years of adolescence. At thirteen or fourteen, in most countries, a change of voice and the appearance of a beard indicate a sudden advance made from an imperfect to a perfectly organized man.

An early maturity characterizes animals generally. Insects have an exceedingly rapid series of evolutions. To be born and become the parents of a numerous offspring in a single day, and then give way to a coming generation, are extraordinary circumstances. Those living longest are the slowest in being physically perfected.

Poets have exhausted their magazines of imagery in their meditations on the helplessness of infancy. But the compensation for those years of incapacity of body and mind for any of the responsibilities of life, is found in the longevity of the race. The average of existence far surpasses the life lease of the general animal kingdom.

This subject will have special consideration in the chapter on longevity.

Whether the moon exerts any more influence on the adult female than on the adult man, is left open for the discussion of professors in their official intercourse with their pupils. Were some learned pundit to assert the planet Mars, rather than the cold moon, the controlling power, who is able to confute it?

There are peculiarities of structure and functions of ourselves, which are divine mysteries. Nature eludes our best concerted efforts for watching processes in her laboratories. Ever vigilant and uncommunicative, we are still profoundly ignorant of what we most desire to know.

We know how life terminates, but who knows how it begins?

# CHAPTER XXVI.

## THEIR MALADIES.

REFINEMENTS are accompanied by a train of discomforts, particularly severe upon women.

Unfortunately, an impression prevails just where it ought not to be entertained, that their organization is so delicate they cannot have exposure to air, exercise, labor, or play, such as men are exposed to without detriment to their health.

As already explained, their anatomical structure is no more complex than that of males. There is a delicacy and a finer finish, if that expression is allowable, but otherwise there is nothing in the form or development of the female body which indicates its unfitness to resist atmospheric changes or any influences from without, which the physical constitution of the other sex can successfully withstand with impunity.

Hardships are met with in all conditions of life. Pleasures and pains are about equally divided. Finally, there is no special reason why women should not be as free from sickness or indisposition from ordinary causes, and endure as much and as long as men, all other things being equal.

Hereditary infirmities, such as scrofula and pulmonary consumption, appertain to some families. Easy circumstances present no reliable modifying conditions that promise less suffering than is to be met with in abodes of poverty.

## CHILDHOOD.

Among the poor a deficiency of proper kinds of nourishment in childhood leads to physical conditions which are troublesome in adult years. Especially so in regard to female children. Restrained as they generally are from out-door exercise, allowed to boys, and restricted most unfortunately, more frequently than otherwise, to small, badly ventilated apartments, their domestic pursuits ordinarily being sewing, or of a kind that keeps them most of the day and all of the night in social imprisonment—they grow up with less firmness of bone and muscle than their brothers, since no one cares whether their faces are tanned by the sun, or their feet are wet by wading in gutters.

Frequent exposure to wind and weather, without reference to temperature or humidity, does very much toward building up a hardy body. Being kept from such influences debilitates, and those thus reared possess feeble powers of resistance. A boy braves the storm, while the girl wilts and fades away under circumstances of home discipline, that robust, stirring, boys could not endure.

On reaching womanhood, a girl is not able to resist influences that destroy her, while young men contend with the same contingencies without being moved by them in respect to health.

When the teeth show defects as early as fourteen to sixteen —not unfrequently much sooner—it is pretty conclusive evidence of the premature death of those organs, resulting from an insufficient supply of phosphate of lime. The bones are not usually so well grown nor so strong in young misses whose teeth exhibit a paucity of that element, necessary for the perfect development of the whole osseous structure.

We have already shown that in agricultural regions, where

cereal grains are easily and abundantly cultivated, the people
are taller and their bones are both larger and stronger than the
bones of those who live where crops are only sparingly raised.

In Western wheat-growing districts the inhabitants are proverbial for white, beautifully-set, sound teeth.  When those
materials, which once gave from forty to fifty bushels to an acre,
have been exhausted by continuous culture, without returning
to the soil an equivalent for what has been drawn from it,
the product dwindles to fiften or twenty bushels.  Then defective teeth begin to appear in young persons.  The third
generation, on the same ground, deteriorate in stature.  Short
men and women, descendants of stalwart parents in the same
locality, would have been as tall and as perfect as their grand-
parents, had they been provided with the amount of phosphate
of lime they received in their youth.

With an increase of population, and diminished products in
a once fertile grain-growing area, the result of long-continued
tillage, the inhabitants begin to seek new homes.  This is the
commencement in this country of removals to new lands farther
off, which are rich.  An improvement in the physical aspect,
and, certainly, in the height of children born in the new locality,
is noticeable when they reach an adult age.

In cities, a change of diet, even though they may have come
from the pure atmosphere of the country, not unfrequently
immensely benefits some persons by the removal.  It is because
their systems are provided with elements necessary for a complete development of their bodies, which their habitual food in
the interior did not furnish.

Change of location is often quite as favorable, in a physical
point of view, as change of position, after sitting for hours on
the same seat.  That law of change, in relation to man and
animals, is recognized in another form, in respect to a rotation

of crops. It is a gross mistake to attempt raising on the same ground, perpetually, one kind of product. There must be alternations, which afford opportunity for Nature to replace, in her own way, elements that are taken away, and then, after awhile, the grain that had exhausted fertility may succeed again.

## Rest.

No people have ever managed a farming interest so philosophically as the Jews, while they observed the requirements of their great lawgiver. Every seventh year the land rested one year. It gave it time simply for garnering up a new store of salts for raising subsequent crops.

In avenues of trade, handicraft, or in the exercise of professions, competition calls into activity parts of the brain which, in rural life, were almost, if not quite, dormant.

The transfer of some of their vitality from the muscles, as in holding a plow, or reaping a field, requiring no vigorous effort of mind, to the busy scenes and stirring enterprises of a great mercantile establishment, brings out intellectual force in country boys they were not conscious of possessing.

By degrees their faculties are systematized; they grasp at great undertakings in commerce, and when the brain has been educated to the new order of things, mental friction subsides, and slender boys become portly gentlemen, bold calculators, intrepid contractors with government, bank directors, and far-seeing financiers.

## Transplantation of Men.

Society is immensely advanced, as a city is, in being recruited from the honest farmer's one-story house.

A transplantation of men and women is as important for the progress of society as the removal of trees from their native soil to ornament public parks.

Choice fruit-trees are invariably taken from a nursery while young, because they acclimate and accommodate themselves to the circumstances of a new locality. .

Old trees cannot be removed so readily. They die sooner than expected. Nothing is gained, either in quality or quantity, by running counter to those general laws which are recognized by the uneducated as violations, when plants and children are rudely handled, or old trees or old persons are expected to do as well in new conditions, as when left to themselves in places where their habits have been established, and their growth matured.

Select boys and girls for removal to new spheres of life, as a shrubbery is chosen, for healthful appearance, vigor, and flexibility. They can then be handled with impunity, and made to develop where they will be both useful, beautiful, and ornamental.

With females, a change of residence, from rural freedom in a country home to a city, is not so satisfactory in all respects, as with boys. Conventionalities in elevated circles keep them under too much restraint for a play of the vital machinery.

When they come to town with impaired health, it is sometimes extremely advantageous to an enfeebled young lady to have the stimulus of a maritime residence ; or in being transported to an inland town, where they escape the humidity of easterly winds, or long, wet, cold springs, that were causes of indisposition in the locality whence they came.

On being established in town, they find it customary, if not necessary, to adopt quite a new mode of life, which, in connection with close dwellings, heated by furnaces, instead of an open fireplace, with a cheerful glow of blazing wood, together with regulations and preparations for the breakfast-table in one dress, for a promenade in another, at the dinner-table in something else, and lastly, for the drawing-room, in still another change of costume; and all those, independently of very formidable and elaborate transformations for the opera, are direct sources of debility, and, certainly, of great fatigue. She must have a remarkable constitution to withstand so many and unrelaxing causes of indisposition.

## Too Much Comfort.

Women break down under too much domestic comfort, sooner than under domestic hardships. Thus, people, whose days and nights are a series of excitements, high living, and irregular hours, scarcely ever number as many years as those who are obliged to contend with poverty and privations. It is among females in the latter condition that extreme longevity is found.

Men fly about in open air, inflating their lungs with refreshing properties, while their beautiful wives and daughters, with pale faces and tallowy complexions, are lounging on sofas, complaining of *ennui*. How many of them fall like promising blossoms before the fruit is set, killed by kindness. Such is too much civilization.

Travelling for health is nothing more nor less than ranging about for vitality, which all the rich can neither find nor purchase, while the poor have it forced upon them through broken panes and cracks in the walls.

They are pitied because their lot is hard. They have no luxuries for their stomachs; no two thousand dollar shawl to protect their white shoulders; no velvet ottomans for their feet; no frescoed apartments to suffocate in, nor down beds for sleeping away life in idleness. But they have what money cannot procure or physicians furnish, viz., rosy cheeks, sound lungs, white teeth, a good appetite, and other requisites for reaching three score and ten without converting their homes into a hospital.

A consumptive diathesis is most commonly transmitted from the mother. Whether induced in them by exposures, hardships, or transmitted to them from ancestors, cannot always be ascertained. If there were no consumptive mothers, however, there would be fewer victims of that frightful malady.

### DISEASED LUNGS.

Nature is always conservative. The effort is invariably to repair, restore impaired parts, and to strengthen where there is weakness.

There is a contest between life and death in cases where the partitions between the air-cells of the lungs are ulcerated, and the function of respiration is, of course, imperfectly performed. Blood sent there does not obtain as much oxygen as the body requires.

Ulcerations extend and pus accumulates till in advanced stages of the disease extensive abscesses are formed, and cavities are distended, with thick, adhesive, offensive fluid. Breathing becomes more impeded and death ensues.

Usually only one lung is involved in the manner described. Were it within the province of surgery, as it probably will be at no very remote period, when there is more confidence in the resources of that great art and less timidity among operators, that half of the chest containing a disorganized lung will be opened for the extraction of the useless, diseased lobe.

When air enters the pleural cavity, the lung collapses in-

stantly. It would be relieved from inflation, and in that way
set at rest. The sound lung on the other side, completely separated by a partition, and in its own pleural box, would sustain
life unaided by its fellow.

Teachers of surgery set forth in frightful array the fatal
effects of an inflammation of the pleura,—the lining membrane
of the thorax,—should air be admitted to it.

Let them devise methods to prevent its access. No domain
of operative surgery is so miserably handled as that of the chest.
No progress has been made there in a hundred years.

There are thousands of medical men who remember the
perpetual caution impressed upon them in their pupilage, not to
wound the peritoneum. Even a puncture was to be avoided
with scrupulous care, because it was a serous tissue.

On account of that bugbear of apprehended fatal inflammation, peritonitis was managed with difficulty. Now, in the extraction of ovarian tumors—nowhere more skilfully performed
than in the United States—but few out of many are lost, and
yet incisions through that membrane are extensive in ovari-
atomy.

Many women have died of those enlargements in past times,
who might have been saved, had there been more accurate
knowledge of what course to pursue in treating the peritoneum.

The late Dr. Mott remarked to a medical gentleman, while
both were observing the progress of an operation involving
parts he was cautioned in his youth to avoid, "Why, they cut
the peritoneum now-a-days as heedlessly as they would cut an
old shoe."

## THE CHEST.

The lining membrane of the chest is also a serous one, performing an office very similar to that which lines the abdominal

cavity—viz., pouring out a fluid for lubricating the organs within.

When that fluid is in excess, the absorbents failing to take it away, the accumulation is a dropsy. To draw it off artificially, an instrument called a trocar is resorted to.

A puncture being made, the fluid is drawn off. The operation is substantially the same in relieving the chest, but not so often performed, as the inflammation apprehended is considered more difficult to control.

There is considerable unoccupied ground in the domain of surgery. The coming operator who has boldness enough—and it will be called daring—to cut into the chest, and take out diseased portions of a diseased lung, will secure great fame.

The right and left lobes open into one common tube; but if the branch pipe on either side were closed, the supply of air would be inhaled as before into the lung whose tube was free.

Not to enlarge further on this subject, deserving as it is of careful consideration, it may be asserted that many persons at this present moment are in vigorous health, who have only one lung.

Gun-shot wounds, bayonet, stiletto, and sabre-thrusts, have often punctured the thorax, and terrific violence to the lungs did not prove fatal.

### A STITCH IN THE SIDE.

In severe pleurisy, adhesions are formed between the surface of the pleura costalis and the one covering the lung. As the lung thus tied begins to inflate, there is a sharp, painful sensation, called *stitch in the side*, which prevents a full inspiration.

When inflammation has subsided, the individual gradually

begins to inhale a larger volume of air. The bridle which held the lung, so that it could not be inflated without pain, gradually elongates, and finally normal breathing is reëstablished.

Females appear more prone to pleurisy, or aggravated inflammation of the lungs, than men. The manner of ligating the waist prevents the descent of the lungs with the fall of the diaphragm to where they ought to have gone in a full inspiration.

Girding the body with stays diminishes the lower end of the thorax. Its capacity is unnaturally small. Long practice has fixed the ribs where they are permanently held.

Such compression deranges the abdominal viscera. The lungs are forced higher up. Chafing, as they do, through a triangular membranous space at the root of the neck, the sharp horizontal edge of the first rib creates inflammation, and that degenerates into something worse. Matter forms, and cell after cell is laden with the accumulating pus.

The mechanical effect of girding the waist has been explained. A full inflation of the lungs keeps gradually forcing the upper part upwardly, till ultimately a portion rises above the level of the first rib.

This is the origin of many a case of consumption, developed by tampering with the body to make it take a form which is contemplated as an improvement. The penalty is a life of suffering to many, and premature death to a majority of all who have been made over in the barbarous manner so much admired by ladies.

Youth and beauty are sacrificed to the demands of a perverted taste. Thousands of brilliant young ladies have been carried to the grave, victims of stays, busks, and unyielding corsets, the real cause of their premature death not being suspected.

### INCIDENTAL INFELICITIES.

There are indispositions of a temporary kind improperly considered as inevitable results of female organization. This not being a treatise on therapeutics, nor aspiring to the province of a physician, no details in regard to medicine-taking are attempted.

If women reflected upon the mission devolving upon them with more earnestness, they could not fail to perceive that they have not been forced into existence to suffer, nor to die prematurely. Their organization has incorporated with it compensating powers of resistance.

If women are the weaker sex, or in any respect inferior to men, the cause of it is a fault of civilization.

Direct causes of functional derangements, out of which grave difficulties arise, are traceable to actual violations of sanitary laws.

Too light clothing, improper food, imperfect nutrition, the wild waywardness of passion, the seductions of fashion, and the pride to look better than they fancy they appear, and striving to improve their form to correspond with an ideal model of exterior perfection,—are each and all of them dangerous, and, when carried too far, eventuate in some form of sickness.

Why should not a woman be clothed as warmly as a man? That question is not a new one.

### TEMPERATURE.

A notion prevails, and perhaps not entirely without reason, that their ordinary temperature is higher than the vital heat of men under precisely the same circumstances.

Admitting it were so, that they have less need of thick

clothing, it fails to explain why they are ever dressed in such frail fabrics as are scarcely sufficient to resist a zephyr. Most of their garments are rapid conductors of caloric.

Mothers cannot be ignorant of this fact, that the clothing of their girls is far lighter and less substantial than that of their boys.

It has been heralded from Dan to Beersheba, in treatises expressly written for the instruction of females, and by warning voices, that the present method of clothing young girls in this fitful climate is wrong.

But it amounts to nothing. There is no improvement. An appalling percentage are doomed to die before they become responsible beings.

More females than males, according to necrological reports, die annually of consumption. Were men subjected to the same stay-discipline from a tender age, their ribs distorted, and their lungs preternaturally operating in a place too small for the oxygenation of the blood, the bills of mortality would exhibit melancholy memorials of the death-rate of the self-sacrificed.

Men die of consumption. When sporadic, and not hereditary, it may be traced to exposures that brought on severe bronchial inflammation, respiratory derangements, and their concomitants.

With wide-spreading ribs at the base of the chest, they resist, successfully, influences which the female chest in its distortions cannot withstand. Therefore their hold of life is less precarious.

Consumption is only one of many diseases to which females are liable, that may be avoided. If, as physicians assert, derangements in the lungs, engorgements and congestions of the mucous membrane of the pelvic viscera, are due oftener than suspected to their insufficient garments, there is a remedy without resort to medicine.

Women expose themselves with thin shoes, and insist they are thick enough. In their thin silks, and other delicate dresses, in going into the open air, they cannot resist the sudden blasts that chill them in passing from one temperature to another.

A few of the many may have the moral courage to be comfortable at the hazard of being represented as eccentric or opinionated, regardless of what people may say of their oddity in not killing themselves in the wake of fashion.

### NOT CLOTHED SUITABLY.

Small girls in the house, the street, the school, and in their amusement, expose too much skin surface to the weather.

Fancy growing boys, wearing summer coats without sleeves in winter, their necks bare and bosoms open to cold breezes, with a postage stamp on their caputs instead of a hat, racing at foot-ball in kid slippers,—and they would convey no inapt idea of the scanty clothing of female children generally in the Atlantic States.

Bare arms, bare chests, light tight-fitting dresses, and, lastly, their shoes and gauze stockings, are their certain destruction.

Girls should be as warmly clad and in as thick clothing as their hardy, red-cheeked brothers.

Discussion is not invited. This statement is presented for the consideration of parents. Either allow girls to exhaust their superabundant vitality in unrestrained out-door rambles, barefooted and bonnetless, like vagrants, which would contribute to robustness and vigor; or dress them suitably for protection against cold blasts that they meet in their pastimes from an overheated parlor to an open piazza.

Immense numbers of young girls are always to be seen in attics barefooted, in clothing that outrages decency, whose cheeks glow with health. Their homes are cheerless; they lodge in rickety apartments where fresh air reaches their lungs through broken windows and unfastened doors. Their food being plain, coarse, and often cold, their digestion is not deranged by high-seasoned dishes, too strong coffee, or their nerves excited beyond a normal condition.

While such children move our sympathy, and the demoralizations to which they are exposed are deplored, they have what wealth cannot purchase,—health. The rich man's daughters pine with their feet on velvet carpets, and they repose on down-beds when their eyes are closed in slumber.

Poor girls are by no means wholly exempt from sickness. There are painful sacrifices of human life in the abodes of poverty. Yet decaying families are recruited from the ranks of those which oftener than otherwise are regarded with contempt by the vulgar rich.

This idea does not embrace the haunts of vice, but simply refers to the country where children generally inherit sound constitutions. Their capital in the future business of life for securing respectability, position, and independence, is made up more of honesty and energy of character than money in bank.

### Family Failures.

In cities, especially those active commercial centres where wealth becomes literally immense, families deteriorate rapidly, and virtually become extinct in about two hundred years, pampered and placed above a necessity for exertion. Such is the progress of decay in a country like this, where no laws of primogeniture secure posterity a foothold on a landed estate.

If it is true that many noble families have been perpetuated in Europe, as their own historians assert, by having plebeian blood incorporated with their own, it is not improbable that a farmer's son, or a chambermaid engrafted upon a withering stock, will save many a name and many estates on this side of the Atlantic.

The ills of women multiply with the progress of social refinement. They are usually traceable to causes that might have been avoided. No revolutions for their special benefit are anticipated which call for an abandonment of customs or etiquette intimately interwoven with the present aspect of civilization.

More courage would be required to stem the current of popular prejudice among those who make up good society, in common parlance, than to subdue a rebellion against the government.

Therefore they are doomed to suffer, in order to be consistent; and they must die prematurely, because it would be unpardonable to live in defiance of the public sentiment in regard to what is deemed to be extremely respectable.

### DOSING TOO MUCH.

There are special infirmities appertaining to women—technically recognized as the better class—so common, and so many are afflicted, that a paragraph or two will be sufficient to open their eyes to impositions practised upon them by unprincipled medical specialists.

Both male and female pseudo-medical practitioners are equally guilty of fraud; and the only possible way of limiting their demoralizing manipulations, which generally aggravate conditions, is to expose their nefarious doings to the indignation of those whose confidence they wickedly abuse.

No one but a physician can estimate the extent and ravages that result from over-treatment of simple local difficulties, that would eventually correct themselves if left to the recuperative efforts of nature.

Young women, quite as often as matrons, present anomalous pelvic complaints. Even a slight congestion is magnified into a bugbear, requiring very special attentions. As the patient relies on the report of the only one consulted in her case, the opportunity for keeping up an alarm is quite within the control of the person consulted.

Since female practitioners have been recognized as being in an appropriate sphere, a woman very naturally gives them a preference. That is proper and commendable, but there are female quacks.

Because a seamstress can increase her income by announcing herself a physician, without the slightest preparation for the responsibilities of the profession, she should not be consulted before exhibiting some honorable evidence of her qualifications.

Very respectable physicians, in most respects, unfortunately for the progress of medical intelligence, have their hobbies.

For the last dozen years a mania for caustic applications for almost any engorgement, or slight inflammation of some mucous membranes accessible to the practitioner, has raged with the intensity of an epidemic. Thousands of women have submitted to a topical application with lunar caustic, who were injured excessively by it.

Because too many, improperly cauterized, have kept the secret of improper treatment to themselves, it is hoped this exposition of an imposition practised upon them, may lead to the correction of an outrageous kind of practice.

Let no woman in her senses submit to the nitrate-of-silver

treatment, without consulting previously the most commanding medical authority to be had.

Two-thirds of the self-styled female physicians, whose signs figure conspicuously in basement windows, are quite as ignorant as self-reliant, and without a ray of pathological knowledge.

When there is cause for alarm, induced, perhaps, by fatigue, or from any other cause, let nature have an opportunity at restoration first.

Nutritious diet, warm baths, a mild course of tonics, wine, new sights, new faces, and breathing another atmosphere, purer and less contaminated with street dust, coal, gas, or other impurities, are superior as curative agencies, and infinitely more to be prized, than a farrago of medications.

Gross impositions are practised on the credulity of sensible women, too, by unprincipled speculators in health.

It is their misfortune often to require advice, which they should have from reliable sources, but it is extraordinary that persons of good understandings are as often duped as those of no intelligence.

One of the reasons why medical gentlemen of reputation hesitate to give countenance to female practitioners is, because there are such impostors among them, unscrupulous cheats, deceiving their own sex whenever opportunity presents; and honest female practitioners have to suffer for their sins.

Moral influences, appropriately directed, should close the gates against medical adventurers. If the people, particularly the female portion of the community, are taught as they should be, in the course of education, the fundamental principles of physiology, they would not be so frequently deceived in matters pertaining to their own health, and by ignoramuses, too, whose ignorance is concealed under the title of doctor.

There are but few positively sound women in this country.

Many are unsound who might have been models of physical perfection.

Causes which tend to disease, and, consequently, to an abridgment of life, have been sufficiently set forth, but with no hope of inducing one in a thousand to abandon their idols.

# CHAPTER XXVII.

## THEIR POWERS OF ENDURANCE.

What Women can do—Under Pressure of Misfortunes—Distinguish Themselves in Science—Being Misplaced—What Offices they could Discharge—Out-door Employments—Capacity—Iceberg Philanthropists—Children of Indigent Parentage—Exposures to Varying Temperatures—Development of Strength.

WITH a delicate organization, women certainly endure bodily sufferings with firmness and heroic resolution.

They can do anything in art or science which the other sex accomplish. Certainly, they have the ability for mastering languages, playing music, or carrying on nice mechanical operations. In sculpture, painting, and many ornamental arts, they vastly excel. If they had fewer muscles, or fewer bones, or even more than a man, they could not conduct manipulations requiring expert fingers and a well-formed brain.

Annals of war furnish thrilling accounts of brilliant achievements in arms, in which young women braved the hardship of a camp, with a fortitude that would have exalted the reputation of a veteran, without shrinking. They have often triumphantly gained a reputation for skill, bravery, and patriotism.

Their capacity for horticultural and general agricultural pursuits, is widely acknowledged. In their poverty and dependence on manual labor for bread, their strength keeps pace with their necessities. Thus, in Europe, they till the soil, drive teams, saw wood in the streets, act as hostlers, and to the disgrace of those communities in which their hard destiny compels them to do the work which belongs to stronger and naturally rougher hands.

It dwarfs them, to be compelled to carry heavy burdens. Exposure to all weathers hardens and tans their complexion, while alternations of heat, cold, rains, and winds, bronze their skin. To be sweating and tugging in the laborious pursuits of a farm, is not their appropriate sphere. Still they do it, and adapt themselves to the hard fortune imposed upon them, without complaining more than the sisterhood whose destiny places them beyond the necessity of being industrious.

They bear up under misfortune, indeed, under all hardships, more cheerfully than men, without perilling their health or morals. Their instincts are always in the right direction.

A mother in the extreme wretchedness of some forms of poverty, ignorant, and dependent, manifests as much maternal solicitude, affection, and unconquerable love for her children, as the wife of a peer. She submits with fortitude to surgical operations, and endures protracted pains more heroically than men, whose physical powers of resistance, apparently, are far superior.

A citation from historical records to establish this proposition would be needless, since it has become a proverb that a woman is acknowledged to bear away the palm.

When circumstances require, women do as well as men as teachers, artists, or bookkeepers; and they are constitutionally more honest than those claiming to be lords over them. Having the same number of nerves, bones, and blood-vessels, why should they not do whatever men do in those economies which require brains and hands?

If they fall below the sterner sex in any sphere of action, it is because their education has been less complete. Give them equal advantages.

Immense numbers of men and women are unfortunately misplaced. Society, consequently, is a loser by not having them in positions where each would have contributed advantageously for the good of all.

Women quite frequently find themselves mismated as well as misplaced. It is a mistake they often make, in supposing that pearls and diamonds are worth more than intelligence. Jewels have no weight in an intellectual balance.

"When unadorned, adorned the most," is a trite expression, but it conveys a truth applicable to women of culture. They have an influence wherever they move, because there is a force in their deportment, and especially in their words, when properly directed, commanding both respect and admiration.

An unaccountable opposition is manifested against granting educational privileges to women.

There are unsuccessful merchants who would have been excellent farmers, and many farmers of the most thriftless order, who would have made enterprizing traders. The pulpit is burdened with stupid clergymen, whose voices are an anodyne, and their reasoning solid opium. Their congregations sleep as quietly under their clerical administration of the parish, as if they had taken a dose of chloroform at the commencement of the services.

Now, such somnambulant church-operators might have succeeded far better in pursuits requiring muscle instead of brain. Lawyers, too, profoundly ignorant of law, and physicians who literally know nothing of the profession they are permitted to practise, are familiar examples of social displacement.

One of the objections to giving women clerkships, or putting them in positions of accountants, actuaries, bookkeepers, bank-tellers, ticket-takers at railroad offices, and even conductors, and many other pursuits, which they might follow quite as accept-

ably as such services are performed by rough, coarse, unmanner-
ed vulgarians, loathed by those obliged to come in contact with
them,—is from a fear they might become demoralized by such
general intercourse with the outside world. Theoretically, but
without any valid reason, they should be occupants of the house
at all times, and the instruments with which they should become
familiarized, are broomsticks, needles, and teapots.

It is discreditable to men, who have so little confidence in
the moral perfections of their mothers and sisters as to exclude
them from situations which would not only be eminently beneficial to themselves, but also to each and every community in
which their fitness and capacity for such pursuits have been
appreciated and encouraged.

## OBJECTIONS URGED.

It has been urged that women could not be relied upon in
some very common offices in which men are in charge, on account of certain constitutional peculiarities, which forbid females
from exposing themselves to varying temperatures, wetting their
feet, etc., which would prove ruinous to their health.

Such apprehensions are unfounded. It is true, that those
reared so tenderly as to make them unnaturally feeble, and
therefore more susceptible, would not have constitutions for
some industries; but a woman who has been allowed through
her childhood to breathe in open air, to exercise her muscles
out-doors, can resist any and all influences from atmospheric exposure, that a male organization resists. Early training, and
not a congenital predisposition, fits or unfits either for activity
and usefulness, in-door or out.

In all discussions on the subject of female suffrage, a sort of
epidemic that breaks out occasionally, to the immense alarm of

politicians—especially those who dread the elevation of women, well knowing their intelligence and superior moral qualifications would be a death-blow to their own aspirations—it is assumed that they are physically unfitted for pursuits in which men engage.

Mental capacity is ingeniously left out of the account as much as possible ; and those very wise doctors who, in fear of having well-informed women employed to nurse sickly or demoralized institutions, are continually harping on their inability, are careful to say nothing about their educational fitness to transact affairs far more successfully than thousands of party numskulls, whose only qualification for positions they disgrace by ignorance, is devotion to leaders bolder and more unscrupulous than themselves.

### WHERE THEY WOULD SUCCEED.

In courts of law ; on grand or traverse juries ; as coroners, sheriffs, and similar offices, which require tact, good manners, firmness, and an accurate knowledge of legal forms, women could aquit themselves far more acceptably than such coarse, profane, offensive occupants as sometimes hold those places.

In March, 1870, an experiment was tried in the Territory of Wyoming, for the first time since the formation of a civil government in America, of placing women on a jury.

A wretch, by the name of Cowie, was on trial for murder. The panel had upon it six females and six males. After a protracted deliberation of four days and nights, a verdict of manslaughter was rendered.

Of course, the ladies were exceedingly fatigued, but their resolution, and the dignity and solemnity of the occasion, won for them the admiration of the whole country.

Hardly, however, had the eclat of their services been heralded over the country, than it was bruited about that their husbands were dissatisfied with such a protracted absence of their wives from domestic duties. Worse still, a busy press was active in propagating a story that another source of dissatisfaction grew out of having their beloved helpmates shut up four days and nights with strange men, sturdy yeomen, of whom they knew neither good nor evil.

That must have been an attempt at merriment, or the outgrowth of a mischievous disposition to destroy the influence which women were acquiring in their praiseworthy efforts in a new and important sphere of action.

### HARDY DISCIPLINE.

Children born to apparent affluence, tenderly managed, by unexpected family reverses have been often thrown upon the cold charity of the world to grapple with poverty in its severest forms. Iceberg philanthropists seldom thaw at the sight of wretchedness that can best be warmed by money.

In transitions from one social extreme to another, the body suffers from no shocks that essentially impairs it, while a hope is entertained of ultimately rising above poverty to the realization of influence and comforts. Some fall by the way, whose feeble organization is unequal to the depressing wretchedness of hope deferred. But how many live through painful scenes of want and mortification to reach old age in a better aspect than when they first began to battle with tribulations!

Children of indigent parentage throng the streets of every city, barefooted, hatless, bonnetless, thinly clad, and oppressed by hunger, braving storms, whose ruddy cheeks bear witness to the invigorating influence of fresh air.

Female children from such sources are factory operatives. They are domestics, and in whatever position they may occupy at service, are not only expected, but are ordered to do that which as often as otherwise obliges them to be standing in water—handling wet clothes, cold and chilling to their warm blood; and yet they sustain a higher standard of health than the pampered offspring of their masters and mistresses, imagined to have been born to a better inheritance.

It is no more dangerous to have one's feet in cold water than to have their hands in it. There is nothing in the anatomy of a woman's body that indicates a greater susceptibility in her feet than in her fingers.

The whole body, as individual limbs, or the face, may be accustomed to endurances that would be detrimental to a novice in such kinds of exposure. A sudden plunge into a cold bath reduces vital temperature. In coming out, it returns with accelerated force. But the after-glow, so much coveted by ladies, and of which they speak with enthusiasm, as delightful in the transition from a bath to a warm room, is a dearly purchased pleasure by some hydropathic advocates.

That after-glow draws largely upon the vitality of those of extremely delicate organization. It takes so much from the fountain, that it finally ceases to rise to its normal level. A lady may dissipate in a bath, to her injury, quite as readily as with chloroform or opium. Their effects, however, are widely different, though both lead to the gates of death. Excess in anything enervates. Regularity, even in the violation of organic laws, does not produce derangements immediately.

Sudden cold douches are as unbearable as electric shocks; still, by gradually practising, as, for example, keeping the hands or feet a long while in intensely cold water, no injury ensues.

Pearl divers descend thirty and forty feet, and walk about

deliberately on the bottom. Suspending respiration thus is an education of the lungs to meet the contingency of their profession. Washer-women in Paris paddle in the Seine with the freedom of ducks—always cold and wet; yet they live to the ordinary age of those who have had no experience in aquatics.

The muscle of men becomes stronger and more massive than in women, because they are more exercised in all forms of activity. Just as the twig is bent, so is the tree inclined.

As soon as muscles are required to perform an increased amount of service in a particular routine of action, an extra volume of blood circulates in them, which is equivalent to giving them more food than when their labors were less.

Thus, a blacksmith's hammer-arm is larger than the other, because the weight to be habitually moved in forging at his anvil requires increased muscular force. It is, therefore, more copiously nourished.

The stonecutter's arm becomes larger that wields a mallet, than its mate directing the chisel. Ballet-dancers, rope-performers, circus-riders, and professional pedestrians, have wonderfully fine lower extremities, while their arms appear disproportionably small, in consequence of not having an increased circulation directed to them.

On the other hand, porters, or those who are constantly handling, moving, and lifting heavy boxes, barrels, etc., or carrying burdens on their backs or shoulders, have a prodigious massiveness of the pectoral muscles about the upper part of the chest and lumbar region.

It is one of the strange sights at the port of Havana to watch the play of muscles of nude burden-carriers in discharging vessels, which stand out in living prominence.

Stevedores, in Sicily, walk up a plank with a bale of rags on their brawny shoulders, weighing, upon an average, four hundred pounds. They move off deliberately, as though not particularly embarrassed.

We have a distinct recollection of seeing a Turkish porter, passing through a street in Smyrna, with a barrel of New-England rum slung to his arched back.

Processes which develop strength in men, will also develop strength in women. Arab girls, on the banks of the Nile, and indeed all over Syria, assist one another in raising heavy jars of water to their head, which they carry off to distant villages with apparent ease, rarely touching the vessel with their hands, so admirably are they poised.

Files of those dark-eyed, supple nymphs, in social chat, cheerfully wend their way for one or two miles, without the least apparent fatigue. Such habitual exercise of all their muscles brings out the finest imaginable proportions of the body. Every fibre is urged to a full state of tension.

Those bronzed-females, whose symmetrical forms cannot be excelled in any country, know nothing of numerous complaints which are the burden of our civilization. They have neither distorted spines, drooping shoulders, or contracted waists. Maternity is rarely attended with anxiety. Apprehending no danger, they are never harassed by nervous anticipations, or depressed with thoughts of danger.

Were orthopedic surgeons, whose specialty is to warp distorted bones into position, to treat their patients to sustaining weights on their heads, and exercise with them, their success would be far more satisfactory. Put the muscles into action, properly directed, and they will certainly adjust the distorted parts, by contracting forcibly till the deviating bone is gradually restored to its natural relations.

## Who are Distorted?

Young ladies brought up in luxurious indolence are the principal sufferers from incurvations of the spine. Family opulence is not unfrequently the destruction of heirs to an estate. Rich girls are made puny, feeble, and lifeless by their dresses, table luxuries, gas-lights when they ought to be in bed, by opera excitements, piano drillings, unventilated apartments, and brain-burning novels!

When very young, they should be permitted to range in loose garments, and be as free as the poor man's daughters. That is the way to form a good constitution. If, however, the misfortune of a spinal curvature overtakes them, let them promenade regularly with as much of a weight on the head as they can carry. Do it in the garden or open field, rather than in a drawing-room. Being lashed down to an inclined plane is an absurdity, and deserves professional condemnation. Liberate their oppressed ribs; give them coarse food, instead of dry toast and tea. Imitate the vigorous girls of Egypt. Theories disappear before facts in orthopedic surgery :

> " *Natura duce* was the text
> Of ancient Hippocrates,
> But we shall lead old Nature next,
> The force of art so great is."

# CHAPTER XXVIII.

## BRAIN FORCE.

Mental Differences—Genius not to be Purchased—Soul—Molecules of Matter Perpetually Re-arranged—Duplication of Organs—Brains Look Alike—A Divine Mystery—Male and Female Brains—No Apparent Anatomical Difference.

No one pretends to question the universal opinion that intellection is manifested through the instrumentality of the brain, a poorly understood organ.

Brain force, that exercise of the will which places humanity at the head, and gives man control over animals, and, in fact, over the whole world, perplexes philosophers as much now as in the earliest periods of philosophical inquiry.

Science affords but little light for conducting investigations which have in view an easy explanation of cerebral functions. That positive something which is a power, exercised by individuals in producing great, or, indeed, any results, is potent, and almost irresistible in its fullest development.

Some are superior to others, because they originate thoughts. Mechanical inventors, those having the faculty of combining complicated motions, resulting in the production of labor-saving machines, or who conceive unique designs, and execute splendid works in art, must have brains intrinsically different from those who are totally incapable of exhibiting new and striking forms of talent.

Poets, writers of exciting fiction—admitted to possess active imaginations—create scenes and circumstances, which are transferred to paper, as the painter does an ideal image to canvas, to delight those who have no similar inspirations. Some, then, have a capacity for enjoying the mental productions of others. They have, too, a skill in searching out beauties, and of detecting faults, without a faculty of originating.

Yet, in a dissection of the brain, the most accomplished anatomist cannot detect the slightest difference in structure. One may exceed the other slightly, perhaps, in weight. But many a genius has had a small head, and thousands of distinguished fools had a brain surprisingly large.

Misers see phantom dollars upon the same philosophical principle that an architect sees in his mind's eye the structure he proposes to erect. Both contemplate an intangible representation, which is copied and made real.

Whether education changes the arrangement of cerebral fibres, requires further investigation. It develops and directs innate powers which otherwise might have remained partially dormant.

A knowledge of reading, writing, and arithmetic, or utter ignorance of those useful branches, is no evidence of an inability to invent or make discoveries of importance to mankind.

## GENIUS.

Genius can neither be bought, sold, nor transferred. It appertains to individuals. Hence they who possess it in an eminent degree, when directed for the advancement of the common good, are regarded as public benefactors. When extraordinary intellectual endowments are wasted in frivolous pursuits, or the envied possessor of rare cerebral gifts fritters away opportunities for enlarging his own orbit and advancing the interests of the community, society says he lived to no purpose, and died without leaving a brilliant memorial of his existence.

A perfectly developed mind depends on a perfectly developed condition of the apparatus by which it is manifested.

A body may be mutilated to an extraordinary degree without at all limiting the range of intellect. A member of the British parliament was born without arms or legs, yet he is a man of clear perceptions and unclouded judgment.

When all distinguishing characteristics of a well-balanced intellect are active and even brilliant, every limb may be amputated, both ears removed, both eyes blinded. the teeth extracted, the tongue severed, and many more terrible mutilations inflicted without essentially impairing intellection, which remains as complete before.

## THE SOUL.

When those material instrumentalities by which mind is manifested are injured or destroyed, then there can be no conscious volitions.

It is argued that the soul is something quite independent and distinct from the machinery of organic life, through the instrumentality of which its essential attributes are manifested.

If food is withheld too long, a debility of the body follows and the mind falters. If the tissues are not supplied with materials for repairing a waste constantly going on in the system, organs cease to operate. Death ensues, and the soul departs.

Our bodies are all the time receiving new materials, and throwing off effete substance that has imparted its vitality. Let this operation be suspended even but for a very brief period, and derangements and death would be inevitable.

Particles received yesterday are ready to be removed to-day, while new ones, just elaborated from food, take their places. Thus life is mechanically sustained.

It is, therefore, morally certain that mind is an independent, intangible something, which exhibits itself through vitalized matter. From whence it came, or where it goes, belongs to the province of revealed religion to elucidate.

## DUPLICATION.

Animals are organized beings, varying in their forms, both externally and internally, according to a specific service they are to perform in the economy of nature. As far as naturalists have carried their investigations, each and every one, including man, are duplicated in their bones, muscles, members, and special nerves of sense.

Two halves, rarely varying much in form, number, or weight, are united to make one symmetrical whole.

Thus there are two brains united, two eyes, two ears, two olfactory cavities, with two sets of nerves alike on the two sides, two kidneys, two arms, two feet, and, in the fœtal state, each jaw was in two pieces.

An obvious advantage in thus duplicating so many parts, is to increase and concentrate force, whether vital or mechanical. Even the heart is double. One heart throws blood into the lungs, while the other propels it through the body. By welding them together, less room was required, and compactness in packing is one of the wonders disclosed by dissection.

In respect to the brain, nerves, and muscles, when one set are out of order, or they can no longer perform their part in the circle of vital movements, thought, volitions, and muscular efforts are solely carried on by the other half, which is unimpaired.

We hear with one ear, see with one eye, chew on one side, taste with half a tongue, secrete with one kidney, locomote on one leg, and do very well with one arm; and in a paralysis of half the body, drag it about for years, while all the powers of life are carried on and regulated by the sound side.

Worms are an aggregation of rings or sections, each of which is almost a distinct individual, having its own breathing orifice, its own ganglions, or nervous centres, equivalent to a brain, and its own independent locomotive apparatus.

Some of the annelides may be cut into pieces, and each one will become a distinct, complete, independent being.

One set of digestive apparatus answers for a double set of organs in all animals, as one boiler is sufficient for a double engine in steam-vessels.

### DISTINCTION WITHOUT A DIFFERENCE.

To all appearance, human brains are alike in structure. One may be larger than another, but it is quite impossible to discriminate a male from a female brain, otherwise than upon the received opinion that the latter is smallest.

On the dissecting-table, the most expert anatomist could not designate the brain of a statesman from that of a scavenger. They are essentially alike, and yet they differ in a manner, while living, which no one has yet been able to explain.

If we were not alike in regard to the number and arrangements of our organs, we could neither think nor act alike. Anatomists, however, discover no difference in the structure or disposition of the brain, nerve, or muscles. Therefore, a great mystery remains unsolved, notwithstanding all that has been taught in elucidation of the laws of life.

## PASSIVE ORGANS.

With eyes, an invisible conscious entity within the brain sees what is transpiring without. It hears with the ears, feels through the nerves, tastes with the tongue, and contracts muscles by a force acting from within. An eye cannot see, or an ear hear. They are completely passive, simply being instruments constructed for conducting to the soul's residence information that could not in any other manner or way be communicated.

Thinkers who exercise their muscles, in-door and out, discreetly, have a longer life than those who are careless in their habits, and sluggish in their movements.

More women are moved by the brain-force of others, than among an equal number of men; but there are female writers whose mental capacity has not been equalled by the other sex in any branch of literature.

Brain-force is a Divine mystery. Its influence is felt, but that is all we know about it. There is no art or device that did not originate in a brain. There, too, conceptions, complex and intricate, may be kept safely for future use, or remain quiescent till the golden bowl is broken at the fountain.

Whatever is fabricated by human hands, must first have existed in the brain—so legibly photographed there, the mind examined the pattern as the work of imitating proceeded till completed.

There being no apparent difference in the brains of the sexes, and experience favoring the opinion there is none, why cannot women do all that men accomplish of value to society? They are entitled in equity to all rights and privileges in the exercise of the talents God has given them, and no opportunity should be omitted, on their part, for exercising that brain-force in all lawful enterprises and pursuits.

# CHAPTER XXIX.

## OVER-WORKING THE BRAIN.

Must be Exercised to be kept in Order—Must have Periods of Rest—Sleep—
A Sound Mind—Predisposing Cause of Madness—Political Friction—
Oriental Calmness—Sudden Death—Avarice.

THERE is a popular notion that the brain may be over-taxed ;
and it is well founded. Those who dwell wholly and continually on one idea,—a perpetual hobby,—injure the organ by
compelling one set of fibres, tubes, or molecules, we know not
which, to be too much and too long exercised without relaxation.

Relaxation is as necessary for the brain as for the muscles.
Alternations of mental action and reasonable repose are necessary in the constitution of humanity.

Those who have exercised the brain pleasurably, through a
long life of industry, have had clearer perceptions, and a higher
order of intelligence, than those whose mental action is irregular : becoming suddenly excited, and then relaxing into
thoughtless indolence, is particularly injurious.

The more the brain is used without abuse, the more complete its functions. It may be injured by concentrating cerebral
force too long, or confining the mind to the consideration of
one problem, or series of cogitations, to the exclusion of other
thoughts, or the intrusion of impressions that might divert the
mind from the order in which the individual is resolved to concentrate his thoughts.

Hence leaders of *isms*, furious reformers, radical agitators, inventors, who dwell long and earnestly on certain mechanical contrivances, as the quadrature of the circle; mad poets, those creatures of imagination, who feel themselves unappreciated, and, therefore, neglected,—become eccentric, and in extreme cases insane, because the brain has had no rest.

### ALTERNATIONS OE LABOR AND REST.

Intervals of relaxation of one set or parts of the thinking apparatus is necessary, while others are operating. The same law governs the organs of digestion. After the stomach has prepared the food received, it passes onward to the alimentary canal. In the meanwhile, it reposes till the next meal is received, thus recuperating in the intervals. Without such opportunities for rest, derangements would inevitably occur. In fact, they do in those who are continually violating the laws of health, by imposing too much service on that badly-treated viscus. Dyspepsia, gastric pains, and chronic inflammations are penalties for gorging the stomach too much, too often, and with materials that bring on direct disease, in an effort to digest what is indigestible. That is forced labor.

An eye must have repose, the heart is perfectly at rest an instant between its pulsations, and beats on, in some bosoms, one hundred years, unimpaired.

Birds sleep at night; reptiles retire to their holes; fishes balance themselves on their pectoral fins in the darkness or aquatic night in slumber. It is thus, while all is quiet, and each and every animal puts itself in a position most favorable for rest, that nervous force re-accumulates for meeting demands that may be made upon the system the coming day.

The brain must sleep, and, in hours of total unconsciousness, if there are no irregularities in the circulation or digestion, regathers that which invigorates it for the waking hour.

Imperfect nutrition of the brain is quite as much the cause of irregular action as being over-taxed with one burden, or a ceaseless devotion to one engrossing theme.

If the liver is diseased, the spleen disordered, the pancreas, scirrhus, or the stomach inadequate to the performance of its ordinary duties, the brain soon becomes impoverished. It is impossible to carry on its appropriate functions on a short allowance.

## A SOUND MIND.

A sound mind is intimately associated with good health, and that is maintained by nutritious food and perfect digestion.

Lunatic asylums furnish painful examples of impaired brains, but those institutions have not yet had the independence to publish such details as would satisfactorily explain many true causes of insanity in a large proportion of their inmates.

It is, perhaps, an exercise of philanthropic discretion not to report what might mortify, pain, or horrify.

A predisposing cause of moodiness, nervous excitability, melancholy, and various phases of insanity, may be traced almost invariably to a violation of some law of life.

An apology may be found for an unfortunate sufferer, by pleading his ignorance; but it is, nevertheless, a transgression. It is charitable to presume hard study has destroyed many promising intellects, but medical authorities teach us that the mind is oftener overthrown by the practice of vices than by an influx of knowledge.

Rather than admit the destruction of reason by intense literary application, writers are beginning to intimate that abuses self-imposed demand a more strict professional scrutiny.

It might be thought premature, or at least inexpedient to announce authoratively in annual reports, that restraints actually bring on madness in some of its saddest forms.

Our civilization imposes barriers against the indulgence of many natural wants. A reflex action deranges the brain.

When Mahometans are insane, it is usually caused by injuries of the skull, frights, sudden surprisals, deprivation of cherished rights, opium, hasheesh, smoking, etc., but rarely, if ever, from moral causes. Moslem fanatics, like those in Christian countries, become eccentric and insane too. Blighted hopes, disappointments in love, or religious fervor, seldom lead to alienations of mind in Orientals.

They have among them fanatical individuals, whose temperaments are like those of the same nervous type in all countries. When thwarted in favorite schemes for revolutionizing a neighborhood or a state, disappointment brings on analagous forms of insanity.

Political rebuffs, unsuccessful enterprises, religious theories which others oppose, self-imposed missions ostensibly for the public good, which were fully intended to be particularly beneficial to themselves, are avenues to lunacy. Each and all of them are proper examples of over-working the brain.

Still, over-working that organ is not quite as common as may have been supposed. A vindictive determination to do what is not agreeable to others, meets with opposition that not unfrequently reacts upon an excited brain beyond what it can bear. That, however, is not to be understood as over-working it. Does a calm, considerate exercise of acquiring, comparing, and analyzing tend to the brain's injury? No.

Have many young men or misses of sixteen ruined their intellect by study?

That their minds have given way in early youth is undeni-

able; but not by schooling the brain in the ordinary manner
of being educated. One-idea people are numerous, and in this
country among women, particularly. When necessity compels
them to laborious devotion to one unvarying pursuit, as stitch-
ing with a needle, running a sewing machine, braiding straw,
reading proof-sheets, or similar exhausting industries, the nervous system is often seriously prostrated. Indeed, the contemplation of one thing all the while, as more prominent than
all others, without reasonable relaxation, is excessively over-
working the brain.

### Explosion of Life.

Commmercial men in communities where property is the
only passport to position, over-work the brain more rapidly
and more frequently than women.

Men occasionally drop dead by a sudden explosion, as it were,
of vital force. Culture, taste, refined sentiments, a delicate
perception of what constitutes good breeding, or lays claim to respect and attentions, weigh nothing where the chink of gold gives
more pleasure than the music of the greatest masters of melody.

Women are apt, with an unexpected change in social position, to become deaf to all sounds not associated with the
rustle of rich dresses, and some die martyrs to an idea that a
wardrobe makes a lady.

Any faculty of the mind may be exercised to its exceeding
detriment. Allowing the powers of intellect to be wholly
given to the acquisition of wealth, to the exclusion of whatever
relates to the moral nature, social duties and obligations, is
avarice. That is, in fact, a disease of the organ in which sentiments are elaborated. It is a malady that destroys the individual before he is ready to enjoy pleasures and advantages he
had promised himself when riches were secured.

A history of trade in its successful aspects, which includes any position in which an adequate income is realized for personal services or skill in the management of funds, would show that not one in five thousand who heap up treasures, ever has the benefit of them. Heirs-at-law, who may never have earned a farthing, usually have the spending of such fortunes.

When a property becomes colossal, a little of it is devised occasionally to eleemosynary institutions, or in special charities for securing the favor of heaven, but not because such spasmodic benevolence arises from a religious sentiment.

It is nothing more nor less than a willingness to purchase what could not be hoped for on the score of merit. A mercantile transaction to the last breath.

To allow avarice to obtain a mastery, is a fatal mistake. The late Mr. George Peabody gave a bright example of the way of finding happiness, by making others so with the abundance which a kind Providence had placed at his disposal. The honored Peter Cooper, of New York, has heaven in advance.

The whole of us, mind and body, must be used, but not abused. Happiness being the object of pursuit, unrecorded miseries are heroically endured to gain what cannot be enjoyed when attained.

Convulsive attempts at reformation, when we are alarmed at a realizing sense of the results of disappointed schemes, is snatching at floating straws.

An over-worked brain must abide the consequences of neglected hygienic laws. For a woman to live many years, she must live simply, industriously, and in obedience to her inborn intuitive sense of what is right and what is wrong, and she must vary her pursuits, so that her brain may have as much opportunity for rest as she requires for her hands and feet.

# CHAPTER XXX.

## Their Complexion.

Physical Bearing—Cosmetics—Let them alone—Eruptions—Pearl Powder—
Water as a Purifier—Pores of the Skin—Temperature of the Body—
Insensible Perspiration—Tint of the Complexion—Antimony.

NEVER perfectly satisfied with what nature in kindness has bestowed upon them, however fresh, healthy, or beautiful, women are continually exercising their fertile minds in pursuit of means for improving their appearance. They tax their ingenuity for increasing the effect of their facial expression and figure.

A man may be massive, bearded, and manifest the highest intellectual power, and yet not be a beauty. Those exterior evidences of his strength and masculine maturity are altogether different from those traits and influences which characterize women. There are concentrated in her person a compound of symmetry, texture, and indefinable properties not readily expressed, which, nevertheless, are felt and acknowledged to exist.

When cultivated, and her soul educated to correspond with her positive corporeal attributes, a woman governs without speaking, and commands by an ineffable magnetism.

She has an innate disposition to appear to the best advantage, and in that way her power is augmented, and her sovereignty over the male sex secured.

## Ambitious to Look Well.

Impressed with a conviction that she can improve her appearance by processes of art, women of all countries are the patrons of cosmetics. The savage female seeks such appliances among simples of the field, and in mineral preparations, which make her hideous, in her fancied metamorphosis for the better; perfectly loathsome, if not horrid, to the eyes of a civilized being.

A woman's complexion, the expression of her eyes, the arrangement of her hair, the size of her hands and her feet, occupy her thoughts too much, if truthfully represented by writers of their own sex. And it is unquestionably true they heroically submit to self-imposed tortures, with an expectation of appearing essentially improved in appearance in the estimation of those with whom they associate.

Not one article in the catalogue of miscalled beautifiers, of which ladies are usually munificent patrons, is worth having, or free from objection on account of deleterious properties in their composition.

Most cosmetics are positively injurious to the skin.

There are no exceptions in favor of any, however popular they may appear from the representations of schooled advertisers, or the opinion of fair customers, to the contrary.

## Cutaneous Blemishes.

Eruptions, cutaneous enlargements, chronic inflammatory flushes, bordering on erysipelatous redness, resisting ordinary discutient applications, are always made worse by such improper treatment as many an indiscreet woman voluntarily imposes upon herself, under a hopeful expectation of a triumphant success in dispersing them.

Women relinquish their idols reluctantly; therefore, the probability of convincing them by arguments, or even the presentation of facts, that they would gain vastly more by abandoning the external application of washes and powders, which they have been accustomed to regard as important appendages of their toilet, is not entertained.

### ROUGEING.

Paints have been found with female mummies in the catacombs of Egypt, with females of an extinct race in South America, and even in the superficial graves of the aborigines, wherever the Indians have resided on this continent.

Bountiful supplies of coloring materials dug up occasionally with the crumbling remains of human bodies, must have been considered indispensable adjuncts to female beauty by those who placed them there, and prove the immense antiquity of such appliances. Some such discoveries antedate the Pentateuch.

On all the continents, but especially in America, revelations from very ancient graves testify to the vanity of the sex, and prove, moreover, that the leading elements of their character have always been the same in every country, in every age and climate, in carrying to their last resting-place materials which were contemplated as necessary in eternity as while sojourning on earth.

Pearl-powder ranks well with ladies, being extensively used by them. A vague notion prevails that it is actually pulverized pearl, and consequently must improve the skin when rubbed upon it.

Such ignorance, however, is only found among very

superficial fashionables, who have no aspirations beyond making a favorable impression, not by words, but through the instrumentality of art.

It may be distasteful intelligence to assure those who pay liberally for genuine pearl-powder, the most approved samples are nothing more nor less than starch. Such as they purchase for their laundries by the pound, for stiffening garments, is just as good and valuable as that sold in quarter-ounce packages at several dollars, under the name of cosmetic pearl-powder.

To be appropriately pearled for street appearance, it is usually dusted on so profusely, as to give the self-satisfied adorable a very mealy look. If some of those pearled promenaders, not unfrequently to be met with, were to dip their faces into a dish of flour, who would be competent to decide that it was not genuine impalpable pearl-powder?

To put it on plentifully, especially under the eyes, round the margin of the temples, and on the cheeks, suggests the idea to a spectator that there may be too much of a good thing.

Even were it true that the application of refined starch were of the slightest use in whitening the skin, there is a reprehensible proneness to run into extremes, which is a kind of abuse, not of a criminal nature.

### SUPERIORITY OF WATER AS A COSMETIC.

The experience of centuries places good, wholesome water at the head of all cosmetics. It is infinitely superior to chemical compounds of druggists, and always has been. No complexions compare with those of young misses who have had no acquaintance with cosmetics. That healthy glow which tints the country girl's cheeks, who, unsophisticated and happily ignorant of the mysteries of a fashionable toilette, can neither be improved by art, nor imitated successfully by science.

On being transferred to a city, a young lady first begins to imitate those whom she supposes to be superior to herself. From that day, her facial deterioration commences. Concentrated food, stronger tea and coffee, and more of it than she had been accustomed to at her rural, happy home; later hours, musical excitations, theatrical spectacles, new exhibitions of the follies and frivolities of fashionable life, stimulate the pulsations of her heart. The brain is overtaxed, and with dancing and phantoms, when day is turned into night and night into scenes of bewildering enchantments, the rose is no longer seen on her fair face. She becomes dyspectic, hectic, yellow, and enfeebled.

With this condition come physicians, pills, phials, plasters for a pain in the side, and a troublesome cough.

Pearl-powder will not bring back the bloom of health, nor rouge, spread thinly with consummate skill over a blanched, sunken feature, recall the lost complexion. Hygeia is discouraged, and takes her departure.

## Structure of the Skin.

The entire surface of the body is pierced by an infinite number of minute openings, known as pores,—the external termination of extremely fine tubes,—or sudorific ducts through which we perspire.

Their inner extremities are coiled up in adipose tissue below the skin. Economy in packing, while being protected in a soft elastic bed, is noticeable in that beautiful arrangement which is equally observable in all other parts of the system.

Through those sweat-tubes, aqueous fluid is exhaled, passing from within to the surface where it escapes, and is immediately lost by evaporation.

When the skin is apparently dry, the escape of fluid is constantly going on; but it is not seen. That is insensible perspiration. If, however, there is any obstruction of the orifices, so that the perspirable fluid cannot make its exit, then there is heat and fever.

If the temperature of the body is raised several degrees in consequence of a quick circulation, the quantity of perspiration becomes augmented. Should the air be at a lower temperature, it is condensed and runs down in streamlets. That is sweating.

Habitual application of substances which clog the emunctories of the skin, and thereby prevent the escape of watery collections gathered in the sudorific tubes, must of course be very injurious.

### DROPSY.

One form of dropsy is an undue collection of fluid in the cellular tissue below the skin. If the free escape is proportioned to the quantity separated from the blood, then the equilibrium of health is maintained.

On the contrary, when not passing off regularly as fast as collected, serum occasionally collects in the abdomen, the chest, or the limbs, which constitutes regional dropsy.

Cosmetics of every kind must very considerably interfere with a free exit of perspiration, as a mechanical obstruction. Were the entire body plastered over with a composition which absolutely prevented the outlet and evaporation from the pores, absolutely necessary in the economy of a living being constituted like ourselves, sad consequences would immediately follow.

On the face, where cosmetics are most freely applied, the pores may be rendered quite useless if not destroyed by them. A dryness, roughness, a sickly hue, and premature wrinkles are the penalty of such attempts to improve upon nature.

### Tampering with Health.

Legislation could not effectually stop the sale of quack medicines. People, not by any means the most intelligent, will have them.

This is a glorious land of liberty, in which every one takes what he likes under the name of remedies. Availing themselves of a national weakness in that direction, ingenious speculators accumulate enormous fortunes by the sale of pills and other nostrums, represented to meet all the contingencies of life, which range themselves in the train of formidable diseases.

Oleaginous compounds, not soap, are probably worse than liquids of a stimulating character rubbed on the skin, because they suddenly close up the pores.   The other generates an inflammation that is slower, but equally detrimental.

Washes, which are announced to have a detergent property, but acting upon the same principle, are dangerous applications.

Simply bathing in pure water is a thousand times superior to the most costly articles for giving and sustaining that soft, delicate complexion which indicates health and vigor.

A better idea of the importance of these sudorific tubes may be formed by this curious anatomical statement, that were it possible to unite them all in one pipe, by joining them end to end, there is enough of them on the surface of an ordinary-sized woman, some have supposed, to extend *two miles !*

Remarkable beauties sometimes appear to have become prematurely old.   Faded beauties wilt rapidly when they begin to

show the sere and yellow leaf.  Were some of those cases investigated scientifically, it might probably be shown that they
hastened an event they dreaded, by tampering with their fine
faces with just such appliances as we have here deprecated.  In
their anxiety to prevent the appearance of deterioration, they
produced prematurely that which they intended to prevent.

## REMOVAL OF BLEMISHES.

A yellowish, sallow-colored skin, which cannot be driven
away, even temporarily, by a flush of surprise, is best treated by
water, which acts beneficially.  Children born of painted or
enamelled mothers, are not robust.  Even their mental powers
are inferior.  They are life-long sufferers in consequence of
maternal folly.

Fluids taken into the stomach percolate to some extent directly
through its walls, making an exit by exosmosis on the surface,
after having traversed through various intervening tissues.

It is by that disposition of a portion of liquids swallowed,
the parts are all kept soft, supple, and in a condition to glide
easily one upon another without friction.

By recollecting that the sudorific tubes are so numerous
that five hundred of them exist in a single square inch, it is no
difficult problem to explain the ready transmission of the fluid
they transmit to the surface.

On the back of the hand and foot there are one thousand
pores to a square inch.  On the sole of the foot and palm of the
hand they reach the amazing number of two thousand seven
hundred in a square inch.

On the surface of the whole body of a woman of ordinary
stature, there cannot be fewer than two millions three hundred
thousand of those emunctories.

It is something to ponder upon, that life, so precious to all, is dependent upon the action of such minute, complicated apparatus.

An excuse has been offered for covering up wrinkles with paste, called medicated enamel, etc., that it is a privilege to repair old bodies externally, as it is to take drugs for counteracting diseases.

If it is right for a dilapidated woman to take tonics for improving her physical condition, it has been argued that it is right and proper to attempt improving their complexion, by staining, frescoing, or other means, according to her standard of taste.

We are not discussing the right or privilege to do just what a woman chooses, as a free agent, but contend that the woman who does it, that is, paints herself, makes an egregious mistake to her personal injury.

Paints, on weather-beaten boards, are to prevent the absorption of moisture, which would hasten their decay. On the living, paints prevent the escape of moisture, a function that cannot be interrupted with impunity.

### SCRUPLES AGAINST ART.

Artificial teeth are not classed with cosmetics, as interfering with vital processes, because they do not in any respect. On the contrary, they are important auxiliaries in preparing food for ready digestion.

Formerly it was considered a sin, by conscientious persons, to resort to appliances of art for securing either comfort or an improved personal appearance. The argument resorted to was this, viz. : When any part or portion of the body has fallen into decay, it is evidently the pleasure of the Being who created us, that we should thus gradually go to pieces, and it is wrong, therefore, to proceed contrary to the divine purpose.

Influenced by such considerations, dentists were violating a great law, and wooden-leg makers, wig-makers, and even oculists in the restoration of the blind to perfect vision, are guilty of the violation of a law equally recognized as the will of our Heavenly Father.

It belongs to the history of the dental profession, that less than seventy years ago many toothless ladies, scarcely able to articulate their hostile feelings in reference to the wicked devices of evil-minded men, who proposed to stud their toothless jaws with beautiful artificial teeth, shrunk back with horror at the idea of having such false appliances.

With a determination not to sin by assuming to be what they are not, physically, artificial arms, glass eyes, india-rubber bosoms—so very common at this particular period—would not be accepted by some conscientious people.

Opinionated, sectarian reformers, who are satisfied that their own narrow views are the express will of our Heavenly Father, kick against the pricks of advancing intelligence, but their efforts are useless. There is no *statu quo* in nature, nor can there be in humanity, without the extinction of intellect, and a moral death of society.

Men and women, with the light of modern science and literature, cannot be kept in swaddling-clothes. Those who are perpetually mourning over the good old times, when they were young, cannot give a retrograde motion to the earth in its orbit, nor arrest the swelling tide of progress.

There is another silly vice to which fashionable ladies are prone, that at least should be exposed, that it may be extensively condemned. It is the application of crude pulverized antimony on the margins of their eyelids, and even spread at the base of the under-lid, giving the hollow below a bluish tint. The object is to increase the brilliancy of their otherwise sparkling optics.

It is unaccountable that it should be supposed by quite sensible women, that a bluish shade of the skin,—a diffused indigo shading at that particular section of the face,—enhances their good looks. No grosser mistake ever quickened their enthusiasm.

That is used largely by Oriental females—the occupants of harems, particularly—for the same purpose. But they are semi-civilized, without souls, according to a popular tradition of ignorant Mahometan proprietors.

Repetitions of antimony or khol make the eyes irritable after a while. They cannot bear the strong light, and a slow form of inflammation attacks the lids.

Their custom of staining their nails, palms of their hands, and even the soles of their feet, with henna, shows their position in the scale of intelligence, and their strict adherence to the customs of their equally ignorant ancestors.

It was in Palestine this relic of remote ages—cosmetics—appears to have been extensively employed. Mrs. Jezebel painted her face. The story of her tragical death, by being thrown from an upper story window, incidentally brought with it the curious fact that she painted her face.

Applying a weak solution of aconite to the corner of the eye, now practised, is intended to enlarge the pupil, and enhance the brilliancy of those organs. A dangerous practice.

There is too much that is unreal. There are reasonable boundaries, beyond which it is dangerous to proceed. Such practices as interfere with the higher range of vital functions, should have appropriate consideration.

One of the latest modern weaknesses that has had an extensive run, has been the passion for blonde hair. To meet the demand, scientific skill has provided a preparation to change chesnut, black, or any other head to look as though it were dyed in a sulphur bath.

Mendicant old women wander through the narrow streets of Damascus with flowing red locks streaming in the wind like bunting from the mainmast of a ship. It is the coveted color with them. Whether they are disposed to think it makes them attractive, we have no means of knowing.

There is no composition, however skilfully prepared, that will compare with pure cold water as a beautifier. It is a perfect solvent for those accumulations over the pores, which are chiefly derived from desquamations of the scarfskin. If it does not readily remove them, it is owing to some mineral elements held in it that give it a quality called hard. Emollient soaps with tepid water is a never-failing success.

Simple warm-water baths, without the addition of cologne, camphor, whiskey, rum, white wine, etc., etc.,—which it is extremly difficult to persuade fashionable ladies are not essential,—or of the slightest utility.

Avoid advertised preparations, however much extolled in certificates from irresponsible sources. They are deceptions. Water is plenty, inodorous, tasteless, colorless, and precisely meets the demands of our nature externally.

# CHAPTER XXXI.

## FEMALE EDUCATION.

What Education is not—New Avenues for Industry must be Opened for Women—Excess of Female Population—They have been Neglected.

.LIBRARIES are burdened with essays on this subject, and there is room for more. Every one who has given attention to it, seems oppressed with new theories and plans, exceedingly important in the estimation of those from whom they emanate. Each writer contemplates his own proposition as the only fitting method for elevating woman to the sphere she was designed to adorn..

Men who never had the honor of having a daughter, and desiccated spinsters who will never be mothers, are those most disposed to contribute copiously to the literature of female education. Neither of them are qualified for guides. It is a matter of profound interest to those who appreciate the importance of educational training, to determine how females should be taught to meet the ever-varying phases of modern society.   .

Education does not mean learning to read and write, working worsted artistically, or playing the piano. Nor should the mind of woman be regarded of such small value as to be put off with indifferent instruction.

Christian civilization should righteously recognize her as man's intellectual equal. A question yet to be decided is, whether she is not also his political equal.

If she has not the same amount of muscular strength, she has the same number of muscles, disposed of in the same manner.

Modern thinkers on the constitution and mission of humanity, tacitly admit an equality of the sexes. That old adage, that man is strong and woman weak, is properly questioned of late. A woman's imaginary pictures of moral worth, virtue, and beauty, are better drawn than those by men.

In language, music, and the fine arts, she is by no means inferior. Her mechanical ingenuity in construction is not unfrequently very surprising. The constructive faculty of woman is far above the level assigned her. Devices displayed in needlework, pottery, sculpture, designs, the actual manufacture of metallic pens, jewelry, timepieces, and the peculiar finish given to watches,—the product of their own hands in this country,—confirm an opinion long entertained, that they are unequalled mechanics, when systematically instructed, as men are taught a handicraft.

A needle is a tool. If they can direct that adroitly, as it is admitted they do, they might, with equal facility, vary their pursuits, and use other instruments just as readily. In watchmaking, particularly, proprietors of great establishments acknowledge their unrivalled skill and delicacy of touch.

Therefore, it must be admitted women can do with their fingers whatever men accomplish. Custom, more arbitrary than laws, has placed them where they are not required to engage in many rough employments, ordinarily considered within the province of men, simply because the dress of the latter gives them greater freedom of motion, favorable for a free, energetic, and speedy exercise of their limbs.

## Practical Instruction.

There are trades and pursuits which women are abundantly able to conduct with advantage to themselves and society; and their education, therefore, should have that practical direction which will qualify them to engage in honorable, remunerative efforts. In-door industries, commonly assigned to females, rarely bring them compensation enough for purchasing decent clothing. They are certainly entitled to something beyond the demands of immediate necessity. An opportunity to acquire more than is needed for the present, in reference to the future, should not be denied them.

Such is the extraordinary activity of the human mind at this particular juncture, there is scarcely a branch of mechanical business, however humble, that is not facilitated and made easier through the inventive genius of man. Machines make shoe-lasts, shoes, boots, ox-yokes, rakes, wheels, gun-stocks, mowers, reapers, ropes, cordage, carpets, cloth, hats: and, in short, what is there needed in the daily affairs of life not made by automatic machinery? Certainly spinning, weaving, carding, reeling, sewing, knitting, and hundreds of other similar operations are wholly accomplished by machines propelled by water, steam, or electricity, as though animated by an intelligent spirit within. Cannon cast solidly are bored of any determined calibre, without personal attention, when once the drill is set in motion.

Even pictures are copied by machinery, and news is sent round the globe in a few minutes, so that everything bears testimony to the resources of genius in the production of many modes of doing what was formerly the product of human hands.

One machine performs the work of hundreds of operatives,

and yet nothing is cheapened, as might reasonably be expected with the facilities of this over-fast age.

When boots and shoes, stockings, cloths, hats, coats, dresses, etc., were slowly fabricated by hand-labor, they were far cheaper than at present.

How can it be explained?

A machine moved by steam-power will now turn out three hundred pairs of ladies' boots in one day, and yet they actually cost more than when a good workman could scarcely make two pairs in a day, using his greatest diligence.

There seems scarcely a limit to what is possible, when men of genius interrogate nature.

Therefore, there is a necessity for opening new avenues for female enterprise. The spheres they have occupied from a remote antiquity are closed to them, in the way of industry, by inventions which wholly supersede them.

Women must have bread and breathing-room, even if the population can be served better and more rapidly than formerly by their busy fingers.

Armies, navies, and the mercantile marine take away vast numbers of men. Women remain at home, and hence they outnumber very largely the males in cities and in the old States of the Union. Their prospect is discouraging for sustaining themselves, unless society accords to them the right to engage in pursuits which were once considered exclusively belonging to men.

There are more women than men in many of the European states and kingdoms, and it is so also in extensive countries of Asia and Africa.

This excess of female population is due entirely to the evil propensities of men: their love for roaming excitement, a belligerent disposition, and the exactions of despotic rulers who control their destiny in many countries.

Most cities on the coast lines of the United States have an excess of females by far outnumbering the male population. Sea-service, the needs of new lands for agricultural laborers far back in the interior; mining operations now extensively carried on in the great mining regions of the West, induce men to leave their native places to better their circumstances, while their wives and sisters and daughters remain at home.

Women cannot submit to the hardships, privations, and demoralizing tendencies of many pursuits which characterize those far-off enterprises. There is a rudeness of manner, and a disregard for conventional forms which belong to cultivated society. Civilization accords to women the expectation of being treated as beings holding a balance of power in those social relations which secure propriety and refinement; and all, in fact, which is good, noble, and morally elevated in any community, forbids they should be exposed to the roughness, rudeness, and hardships of gold-digging researches.

### LAW OF EQUALIZATION.

An equalization of the sexes is maintained with peculiar regularity in the animal kingdom. Where there is an apparent excess of one or the other, it is due to local causes; but it in no way effects a law which secures results most beneficial to the perpetuity of a species. There is neither failure in the law of reproduction to meet losses, nor the least danger of extinction, unless a ruthless war of extermination is waged by man, in the hunting of beaver, buffaloes, and whales.

When males are too numerous, they fight among themselves, and slaughter one another till a proper proportion in reference to the females is established. If females are in excess, there is a law of adjustment immediately brought into operation which reduces them gradually without producing violent commotion or perceptible disturbance.

Again, it is equally curious to observe that when there are too many inhabitants in a given area, among wild animals or even aquatic beings, so that the products of the soil or a feeding region of the sea are inadequate to their healthy support, disease comes in the character of an equalizing agent. Thus epidemics and plagues in over-stocked cities invariably subside at a point that saves a remnant, since extinction is not contemplated in a law which the philosopher recognizes as a means of securing a connecting link in a long chain of existence, the loss of which might lead to conditions and revolutions quite beyond our comprehension.

Alarms are occasionally sounded in village lecture-rooms, that women so much outnumber men in the New England States,—being regarded as non-producers in an agricultural sense,—that something must be done to meet the emergency.

It is not alleged they are idle, or in any respect a burden to the community. They consume food, to be sure, and it is equally true they neither plough, chop wood, or labor in the field, nor should they do either.

## Women are Orderly.

There is not the slightest ground for alarm, because women never band together for political agitation; they never prepare revolutions, nor is social order outraged by them, however erratic a few peculiar individuals may appear in vain attempts and exhibitions not in accordance with their nature.

Women neither infest bar-rooms, loiter away the day in saloons, lager-beer vaults, or march through town in hostile bands, destroying printing-offices, or combine in squads for robbing railroad trains. Neither do they stuff ballot-boxes, nor break open prisons for the liberation of thieves or accomplices in wickedness.

They are not proper persons for running up and down the rigging of a vessel. They could not conveniently glide to the extremity of a yard-arm and take in sail in a gale of wind. Their organization unfits them for balancing themselves on a spar while their hands were belaying the wings of a scudding ship. They could not swing an axe in felling forests, drilling rocks in excavating canals, because the management of the instruments used in such labors would interfere with the health of organs essential to maternity.

## WHAT THEY CAN DO.

But they possess all the requisite physical and intellectual qualifications for managing mercantile business, and for sustaining themselves with dignity and success as teachers, from a common school to chairs in universities.

Wherever intelligence, diligence, accuracy, and honesty are in estimation as pre-requisites for positions, women are prepared for them.

They have not been taken into favor in the past, in such relations, because the necessity for it did not apparently exist, as it now does. One sewing machine is equal to one hundred hand-sewers. Yet while they kept all people clothed by their needle industry, their wages were shamefully undervalued.

While their hardy, bold, adventurous fathers, brothers, and husbands are wending their way to distant regions in search of localities in which their prospects would be more satisfactory, their daughters and wives remain where they were, it being neither proper nor always convenient to go with them to border settlements before some preparation is made for their reception.

Women, even in nominally Christian countries, have been so long excluded and neglected, and, worse still, taught to believe it was wrong to be seen or heard outside the house, it has become a prevalent opinion among ignoramuses they ought to remain, there, even if left in ignorance and poverty.

While the idea is nursed that it is improper for women to be exposed to sunshine, because it might bronze their complexion; or exposed to out-door air, they might take cold; or seen where men congregate to buy, sell, and get gain, inalienable rights are denied them,—they are wronged.

What is the duty of society, now that competition in all departments of business makes them far more dependent than formerly,—especially since they outnumber the male population in the great centres of human activity?

### Legislation for Ameliorating their Condition.

Legislation in their behalf practically amounts to nothing. Acts defining their hours of labor in factories or milliners' shops, are farces. It is about the same in respect to the schooling of young girls employed in manufacturing establishments.

They should have both protection and assistance. The latter is the urgent demand.

Ladies of fortune, and indeed those who are amply provided for through an affectionate forecast of provident fathers, mothers, and relatives, cannot comprehend or understand the cry that reaches to heaven for millions of poor, heart-aching, penniless women; nor do those whose beauty has won for them privileges, comforts, and influence which wealth commands, sympathize sufficiently with the less fortunate of their sex who are apparently born to a hapless destiny. Those who are floating on a summer sea of prosperity are especially besought to listen to a plea for help from an oppressed, neglected sisterhood.

There are not agriculturists enough in this country; and, consequently, with an abundance of the best and most productive land on the globe, all the necessaries of life are excessively dear. The supply is not equal to the demand. Western grain-growing prairies might furnish the world, were they all tilled.

## Men out of Place.

Thousands of puny, pale-faced, feminine, sickly, poorly-developed young men, defective in muscular energy, enough in number for a great army, even were half of them mustered in a body, abound in cities, who would have the strength and character of men if they were cultivating land instead of measuring tape with a yard-stick.

They are wasting the best years of life, deteriorating bodily and mentally in counting-houses, banks, insurance offices, confined retail shops, telegraph stations, etc., who ought to be infinitely more useful were they transferred to the open fields, devoted to agriculture.

They should yield their places at desks and behind counters to women, qualified to do all they do, who are suffering for employments for which they are abundantly qualified.

A social revolution is required to purify the corrupt atmosphere of cities, by driving out worthless, dissipated young men, and giving their places to worthy young women. How many delicate stomachs are scantily supplied, and lungs destroyed for want of wholesome air to breathe, boxed up in lofts and stifled apartments, who would be excellent clerks and accountants. Let some philanthropist set the example of patronizing honest females instead of fast moral nuisances.

If those puny, sallow, spindle-legged exquisites, whose greatest achievement is raising a moustache, were to change the society of inkstands for broad acres in the West, they would expand as much in mind as body, and, perhaps, lay a foundation for comfort, independence, and longevity, which are not within their grasp in the confined circumstances to which their vocation limits them, especially when they riot in dissipations.

Those feeble, sickly, neglected girls, in pestiferous lanes, narrow, dark streets, sunless houses, upstairs in sombre rooms, or cellar, should be assisted as they might be, and instructed to command better compensation for their services.

Were loud-mouthed philanthropists more familiar with the painful condition of thousands of young women who might be elevated and directed in useful, remunerative pursuits, by half the attention bestowed upon institutions which do far more for those who have immediate charge of them than for their inmates, heaven would bless their efforts.

## How to Proceed.

First, qualify those neglected girls by sending them to commercial schools to learn bookkeeping; have them taught telegraphy, how to conduct business in life-insurance offices, to be tellers in banks, accountants, designers, engravers, teachers of languages, musical instructors, have them taught the science of surveying; and, finally, qualify them for positions always presenting, where they could do all that young men do in such relations as are indicated in this general scheme for usefulness, and even many more that might be particularized in this miscellaneous grouping of industries.

Young girls, thus qualified, would sustain themselves with honor. And it will be conceded, they are far less predisposed to deteriorating vices than young men.

They neither smoke, drink, nor gamble, visit race-courses, organize boat-clubs, carouse through the night, or engage in any of those dissipations which lead to deceptions, breach of confidence, or expose them to the attacks of knaves or thieves

Defalcations, absconding with funds of a patron, embezzlement of money in their care, forging notes, falsifying checks, etc., would not occur, as they now do, were young women placed where they should be introduced. Their instincts and tendencies, even with no moral training, are always superior to men of the same social grade. They are naturally virtuous, honest, and sincere.

Wherever a pen is in requisition, careful reckoning, exact computation, or an orderly attendance is an element of importance, a well-instructed woman is always equal, and in many trying circumstances, even superior to a man.

It would be a splendid recognition of female ability to sustain responsible positions, were trustees of estates, directors, and other governing spirits in moneyed institutions, to exchange platoons of burly, rough, unpolished, uncivil, bewhiskered clerks, whose thoughts are more on whiskey and tobacco than on the interest of their employers, for an equal number of quiet, delicate, modest, neatly attired young women. They are much more deserving than any one imagines, who simply feels a woman is a sort of a fifth wheel of a coach, only to be cared for when it is impossible to do without her.

They would be less expensive as clerks, and, as experience

would prove, perfectly reliable. Let those who are stockholders, and, indeed, any and all who would encourage the deserving, make the experiment. Their cash would be in safe keeping instead of being squandered in stock-jobbing speculations—so frequently practised by men anxious to turn another's penny in haste to be rich.

## COMPENSATION.

Women should be paid for what they do as much as is given to men for the same service. If they accomplish just as much in a given time, and as satisfactorily as a being in pantaloons, why should they not have the same compensation?

It is disgraceful meanness for an employer to pay only one dollar to a woman, *because* the is a woman, for work in a printing-office, for example, for which a man gets three or four for precisely the same labor, just because he belongs to the masculine gender.

A lame excuse for such unjust recompense is, that the clothing of females is less expensive than male garments,— and further, custom sanctions the scale of prices for labor. But both are frivolous and absurd apologies for doing unjustly.

Whether their clothing costs less or more, is nothing to the point. They are justly entitled to what they earn. Their stomachs are as keen for a beefsteak as their competitors' in full beards, who squander more in one evening at a bar-room than a female compositor could earn in a week at the present rate of compensation.

The chart of female employments has been under consideration for years. Excellent speeches have also been made, beautiful expressions have gone forth, redounding more to the praise of those that uttered them, than to the profit of those in whose behalf they were sent abroad. The poor, hard-working, poorly-paid girls have no more pudding than when nobody cared whether they lived in wretchedness, or died in a hovel.

Political equality and political suffrage for women, perpetually discussed topics with those who make capital for themselves, under a pretence of being oppressed by the wrongs of women, have not yet bettered the condition of the class for whom their sound, but not their substance, has been given.

Political hypocrites and professional philanthropists are leeches, subsisting on what they get out of the people by exciting their sympathy.

After ages may regulate conflicting claims, and settle difficult problems in regard to labor, but it will be a long while before the poor will be made happy by philanthropic resolutions at anniversary meetings, where there are vice-presidents enough to freight a steamboat, but no substantial assistance for the ostensible objects of their overflowing benevolence of words.

We are contemplating the present period; but when the cry of the oppressed goes up to the court of Heaven, where records are truly kept, the claims of that large class, whose misfortunes are the text in this sermon, will be adjudicated, and their wrongs righted.

No objections are entertained against any system of instruction which enlarges the domain of female knowledge, or that qualifies them to act in any capacity in which men ought not to act, while there is an excess of female population.

Parents are bound to pursue a course, in the education of daughters, that promises best for their success in honorable industry.

## What Parents should Do.

There is neither radicalism nor sectarianism in this. When fathers and mothers cannot lay aside property for their children, in this land of free schools, they can qualify them to provide for themselves.

Gloomy pictures might be drawn, illustrative of the degradation of women in over-crowded cities, and the vicious lives some are forced to lead, or die of starvation, from which they would joyfully escape if they could. Life or death are solemn sounds to a shrinking, timid girl, fashioned in the form of an angel, famished in the sight of plenty of which she canno*. taste.

Police courts, jails, penitentiaries, and reformatories present sickening statistics of perverted powers, and wrecks of beauty in sloughs of despondency, that could have been saved to adorn society, had they been cared for by those who, from their position, might and should have taken them by the hand. But it is too much of a sacrifice for some exceedingly good persons to step out of their way to save a saint.

Books need not be consulted, bloody tragedies cited, personal narrations, or painful scenes of misery sought, to strengthen the appeal we are making.

In pagan and Mohammedan countries women have no such unhappiness as is admitted to be common in Christian lands. They have homes in harems which are sacred, under the protection of brutes in the form of men ; but they are never outcasts on the street, seeking like starved beasts of prey whom they may devour.

We speak of them as pitiable objects, ignorant of their rights as human beings to equal privileges, and the same social status, exclusively in the possession of their proprietors, for they are contemplated as property.

With such degradation, however, there are no brothels,— none of that wickedness which is a reproach to civilization, and a curse where women are denied those rights which flow from fountains of justice.

We beg to urge upon those who may begin to reflect anew upon this subject, to assist according to their pecuniary ability in qualifying intelligent young women for something that will bring them a proper and just reward for their industry, more than what they can earn with a needle.

Give them opportunities for acquiring French, German, Spanish, and other languages, and assist them to positions in telegraph stations, where they could make those languages of the first importance for business correspondence. They ought to be, and it is believed they would prove the best, most accurate, and always punctual operators.

Boston, Portland, Hartford, New York, Philadelphia, Baltimore, Washington, Cincinnati, St. Louis, New Orleans, etc., etc., would give ample employment for thousands of such accomplished telegraphers as they might be, if public sentiment were enlisted in their favor.

By opening such avenues as have thus far been closed to them, and by it, virtually compelling young men to enter upon more appropriate pursuits than weighing out tea by the pound, or selling pins and needles, a gratifying change would come over the land. Bread would be cheaper.

We are hoping that phonography, telegraphy, drawing, designing, engraving, and many other useful arts, may be taught in all well-conducted country schools, expressly for qualifying girls in those remunerative branches of industry.

Give young women who may be dependent on their personal efforts, a knowledge of the art or science for which they have a decided preference. If philanthropists will give their support in that direction, health, happiness, and independence will crown their efforts.

# CHAPTER XXXII.

## ACQUIRING LANGUAGES.

Capacity for Certain Pursuits—Waste of Life—Foreign Dialects—We are Called a One-tongued People—How to Acquire a Language—Dogs Learn the Meaning of Words—Curious Facts—Qualifications for Telegraphing.

THERE are persons who have a faculty for making more rapid progress than others in mastering a new language. It is familiar to those wholly ignorant of the science of phrenology, that there is a singular difference among persons of the same age, position, and opportunities, in acquiring specific or general knowledge.

It would be ridiculous to assert that one boy may become just as expert as another in figures or some kind of handicraft, under precisely the same instruction. One will learn Latin or French rapidly, which his companion at the same desk, with the same facilities, cannot acquire so readily under precisely the same training.

Some have an intuitive perception, where others, of equal intelligence, cannot make satisfactory progress.

There are natural mathematicians, as there are, also, natural linguists. Memory is differently manifested, since some persons remember certain things better than others. One cannot recall names of places or men, yet there is a distinct recollection of faces and peculiarities of each.

By this curious difference in the arrangement of cerebral

matter, there is a man and woman precisely fitted for every imaginable place.

That peculiarity is acknowledged at the temple-door of philosophy. Thus every shade of mental development is recognized, and, as no two are alike, there is a brain to meet every condition and all circumstances in the management of a world.

There is an unquestionable difference in the structure of human brains. Though apparently alike in their general configuration, in the materials of which they are formed, and in the manner, too, of circulating the blood through the mass, the arrangement of the atoms of which the cerebrum is constructed, is infinitely varied.

Different races of men differ essentially in mental force. Size, of course, has to be considered in a search for a reason why one brain is more powerful in resources than another. There are walking polyglots, but far more are incapable of speaking their mother tongue grammatically.

If certain convolutions of the brain are more prominent than others, according to the teachings of those who know nothing about the subject, they are charged with force corresponding to their development. About thirty protuberances are marked on charts, which the disciples of Gall and Spurzheim recognized as locations of distinct faculties.

If a ganglionic elevation happens to be the organ of language, and has been better nourished, or rather more frequently excited than its neighbors, it will give evidence of its superiority; while twenty-nine are feeble or embryotic.

We need not consult authors to facilitate progress in acquiring a new language. Men, women, and children are constantly met who have the faculty of articulating many languages fluently, who can neither read nor write. It is curious that remarkable linguistic scholars rarely contribute much to the fund of general literature.

Certain conditions are thought to be essential in learning to speak a new language; but experience and observation show, very positively, that under even unfavorable circumstances, as they would be estimated by scholars, little children on the frontiers of Germany, France, Poland, Russia, Italy, etc., where there is a meeting, as it were, of strange tongues, are perfectly fluent in four, five, and even in six modern languages, yet wholly unable to read or write either of them.

There is no marvel in all this. They are so located that they cannot avoid having their ears saluted quite as frequently with foreign words as their own. Neither effort, study, exercises, or recitations, are ever brought to their assistance.

## EDUCATION OF THE EAR.

The ear is an avenue through which linguistic development is accomplished, and not by the study of books, or recitations of authorized lessons of grammarians.

Such is the commercial intercourse of one part of the business world with another, there is a positive necessity for one language, at least, besides our own. Great transactions with foreign nations could not be conducted with any kind of facility without the assistance of those who understand the meaning and intentions of both parties, if neither understood the language of the other.

Progress in literature, science, art, and mechanism, would be extremely circumscribed, and confined to narrow boundaries, were it not for scholars who change one language into another, and thus put readers in communication with all mankind.

The people of the United States are regarded as a one-tongued population. Millions of foreigners, representatives of every power in Europe, are interspersed through the land, but they are compelled to acquire English, or pass a lonely pilgrimage without conversational intercourse.

We rarely give ourselves the trouble to learn the language of new-comers from abroad. If they desire a social acquaintance, they must blunder on for several years in order to pronounce the *shibboleth* aright. They are under the painful necessity of dropping their mother tongue. We, on the contrary, give ourselves no concern in regard to their embarrassment in attempting to comprehend or articulate what is so familiar to ourselves. If they ultimately succeed in gaining an imperfect command of new words, it is about all those who constitute the majority of Germans, Danes, Swedes, French, etc., achieve. Where enough of any of them happen to constitute a little community of their own, then they hardly give themselves any anxiety about mastering the elements of English.

Several communities thus constituted have so multiplied at the West, that their schools, and even newspapers, are conducted in their native language. This circumstance has obliged legislatures in several States to publish their laws in several languages, that they may not be in ignorance of the way their rights and privileges are maintained and secured.

## More than one Language.

Educational preparation for the active scenes of life, for which youth ought to be qualified, should include a conversational knowledge of the most important living languages. French and German have the first claim. With these and English, we can hold intimate intercourse with about all Europe.

It is a sad waste of precious hours of a college student's life, to be drilling years before and after entering the institution, in languages which are dead. They are accomplishments, but not necessities. One is a language that has not been spoken for nearly two thousand years, nor will it ever be revived.

No objection is offered to their perpetuity.

One or two terms in the course of a college residence, devoted to living languages,—taught so that they could be spoken fluently,—would be of incalculable importance to the individual in all the after years of life. Latin and Greek are drilled into boys for one or two years at a cost far exceeding the expense of teachers of a living language, to qualify them for passing an examination to become freshmen. After that crisis has passed, very little does any one care for Greek or Latin, unless they are designed for instructors in these departments. Therefore, the expense and time are deplorable losses.

## VALUE OF LIVING LANGUAGES.

It is becoming a question of interest among distinguished writers on education, whether some revolution is not required in elementary preparation for the world in which we live, of more value to the pupil than Greek and Latin — would not living languages be more useful than obsolete ones?

A finishing process in a young lady's education is music and French. Schools exist on a reputation for polishing misses in those two much-prized accomplishments. Not to be supposed familiar with both, would be equivalent to being very imperfectly educated.

Young ladies, presumed to have had the best advantages, are usually taught French by instructors who cannot articulate a sentence that would not shock a French tailor. Session after session at an expensively fashionable boarding-school, they are supposed to acquire exact and, indeed, intimate knowledge of idiomatic French; but not one in fifty knows anything about it beyond reading understandingly to themselves. They cannot pronounce it; nor dare the best of them hazard the experiment in the presence of a French chambermaid for fear of exciting ridicule.

As that language is too generally taught in female educational institutions throughout the country, it is a lamentable loss of time for pupils. The teacher as often as otherwise is a lady not much further advanced in the mysteries of accent than those she is drilling.

## WHERE TO LEARN LANGUAGES.

Foreign languages are taught in cities very acceptably by those who come from Europe, who speak their native language far better than those who have acquired them second-hand through professors who could not make themselves understood in a baker's shop in any tongue but their own, were they starving for a slice of bread.

Indulgent parents expend money freely for their daughters, —but it is a poor method of giving them a conversational familiarity with French,—in country boarding-schools.

Instead of keeping a young lady in school for that particular accomplishment, French, German or Italian, place her at once in a respectable, cultivated French family, or with a German or Italian household.

For example, place a young miss in a family at Montreal or Quebec, in which French is spoken exclusively. Or, if the expense is no object, send her to France. No instruction would be required, unless it were particularly desirable to hasten the process. Simply being in a family is sufficient.

There need be neither plodding in primary books, recitations or any instruction whatever. In an incredibly short time her ears would become familiarized to new sounds. Children are more ready than adults, placed under such circumstances, in acquiring the meaning and accent of words.

While boys are drudging at tasks in Greek and Latin, not essential, they could have a complete acquaintance with two or three living languages, of infinitely more value to them.

In all after periods of life one or two languages in addition to their own would be a thousand times more important to them than a critical familiarity with the orations of Cicero.

There would be less brain labor in this method than imperfectly understanding ancient classics, and boys would be qualified for sustaining commercial relations all over the world, while the best Latinist in Christendom could not buy a paper of pins, were he to ask for them in that scholarly language.

## FOREIGN OFFICIALS.

Our ambassadors, consuls, and commercial agents, sent abroad to represent the dignity of this government, protect our citizens and their interests, have often been the laughing-stock of those among whom they resided, on account of their stupid ignorance of all language but their own, which they not unfrequently barbarously murder.

American consuls have sometimes been spoken of as so illiterate as to be incapable of speaking English grammatically. Their appointment have not always been on account of eminent qualifications. If it is true political services are ever paid for in that way, as compensation for aiding in the election of a rampant partisan to Congress, it is high time the Civil Service Law should be enforced.

Government makes a mortifying mistake in ever commissioning foreigners to consular stations. When the native stock has been exhausted, there will be a reasonable excuse for craving the assistance of men who were never on this continent who do not understand its usages or laws. Travelers cannot conceal their disgust at this kind of patronage. A profuse exhibition of brass buttons, with the stars and stripes waving over empty heads on the shoulders of official nobodies, who would not be invited to dine with a cobbler in any country, are no credit to a great nation of freemen.

Every consul should be qualified and write the language of the country where he is stationed. All the higher grades of official representatives should be educationally qualified for their positions. Ministers plenipotentiary at the principal courts with which we hold intimate diplomatic relations, have not been a whit better qualified than their servants in many instances. Clerks and *attachès* have transacted all business, while the great man takes the salary and does the official dining, to be laughed at behind his back.

## TEACHING FRENCH AND GERMAN IN SCHOOLS.

Were French and German regularly and systematically taught, like other more common but necessary studies in district schools for both sexes, the national character would stand on a higher level, whenever those who have had educational advantages in them are required to serve where such languages are spoken.

Every faculty of the mind should be cultivated, and no one of them permitted to lie dormant in these stirring times, when knowledge is power.

Strange faces, new institutions, and new places, differing from those to which we are accustomed, make vivid impressions at first, but they gradually become familiar. So it is in respect to a new language. It sounds harshly, and may be difficult to articulate, is fearfully guttural, or, perhaps, worse to comprehend, even under the best facilities for instruction; but, as the ear begins to be less severely taxed in catching the new vibrations, difficulties melt away.

When a few words are understood, the way is soon made easier for more. By and by a sentence can be articulated. Before half the anticipated obstacles have been overcome, we begin to chat readily.

Germans, Swedes, Danes, Poles, Frenchmen, etc., who did not know a word of English when they landed in America, soon acquire it sufficiently for all the practical purposes of business and social intercourse. We, however, rarely learn anything of their language in their way of learning.

Little children are delightful assistants when a person is under lingual discipline. They prattle away perpetually, unhesitatingly, and, therefore, give important aid to a beginner. A family in which there are small children should have a decided preference over one where there are none, in selecting a home, where the main object is to be within the hearing of the language which it is proposed to acquire.

Adults are very reserved, fearing to speak, lest they should subject themselves to the critical observations of those who might make merry over their blunders. They hesitate to ask questions when they very much desire to do so, for fear of being considered troublesome, or particularly stupid.

No person, especially those quite young, could be in a family of French or Germans, for example, six months, and not make considerable progress, without having had a single lesson given them.

Indian prisoners at the West, wandering about at the mercy of their savage captors, very soon begin to comprehend the meaning of their uncouth gutturals, and discover their views in regard to their cruel intentions.

Accuracy of expression could not be expected, certainly not attained, without considerable practice, since perfection of articulation must result from long practice.

Children learn to speak without instruction. What can they know of the laws of syntax? They lisp their crude thoughts with charming freedom, years before they are taught the elements of grammar. That appropriately comes into play as they approach their teens.

### Language of Animals.

Dogs certainly understand the import of words, or they could not so readily obey their masters. Their capacity for language is far above that of cats. Puss must see a morsel of meat or a cup of milk, to gain her friendly attentions. She may be frightened, but not with harsh expressions, unless accompanied by muscular gesticulations, when away she runs from impending danger.

Dogs, on the contrary, possess a higher cerebral development. They often acquire a general knowledge of two and three languages, according to their advantages. It is not to be supposed there is any particular effort on their part, or ambition, to remember the exact sense of an articulate sound.

A repetition of words and sentences, as heard in the family where different languages are habitually spoken, ultimately fixes an impression. Finally, they associate certain acts with certain words or commands.

There are dogs almost everywhere, trained to carry and bring letters from the post-office, visit the market with a basket for the family dinner, stand guard through the night, and carry notes at the bidding of the proprietor. They actually learn to obey different members of the household who may direct or command them in French, German, Spanish, etc., as they are spoken indifferently in the establishment.

Not long since a pet dog was brought to New York from Naples, whose intelligence was extraordinary. When directed to engage in certain performances, in Italian, he promptly obeyed; but when addressed in English, the poor fellow was amazingly perplexed. After a while it was apparent he had mastered the meaning of a new language, to a certain extent, which was manifested with signs of gratification.

Vulgar dogs, like parrots, gain a knowledge of a few slang phrases, becoming embarrassed when addressed in terms unconnected with towering expletives. Donkeys and mules, usually regarded as the embodiment of stupidity, evince a nice perception of articulate sounds.

There are many places in Louisiana where those shabbily-treated animals in harness readily obey orders given in three different languages, as either happens to be used by their drivers. Those dumb beasts prick up their long ears in surprise, evidently indicating perplexity, when a strange man takes the reins, speaking a new language to them.

Even oxen, dull and unobserving as they seem to be, listen attentively to what is said particularly to themselves while in the yoke. *Haw* and *gee*, equivalent to right and left, are as perfectly understood by trained oxen as by the teamster. *Back* is another command which an infant might pronounce with equal certainty of having it executed. Horses in bakers' carts, market nags, and milkmen's teams, not only know precisely what the driver bids them do, but they also know the exact residence of customers.

A Vermont farmer is reputed to have borrowed of a French neighbor, across the line, the use of an ox for a day, to take the place of a sick one. The stranger from Her Majesty's dominion could not respond to the bidding of the Yankee, because he could not understand him, although evincing a perfect willingness to pull or turn as his mate indicated. With the best intentions, however, the unmated cattle were constantly committing blunders, to the dismay of the citizen of the republic. Towards evening, in crossing a railroad track as a train was approaching, they were urged to make haste by boisterous vociferations, which quickened the speed of the Vermont ox, but the other, not understanding the loud tones, gazed about with glaring eyes, in view of impending destruction, not knowing which way to move, and, in that instant of hesitation and doubt, was crushed to death by the locomotive, and thus died dramatically, in consequence of not knowing the English language.

Address any of the domesticated animals, accustomed to the sound of human voices, and there is no doubt respecting the fact that they attach a meaning to what they hear, to a limited extent. That is particularly noticeable in menageries. There they become cognizant of the expressions of their keepers.

They learn their ways, analyze their character for kindness, and govern themselves accordingly. So accurately do dogs and cats discriminate a good disposition from a morose, severe, unsympathetic person, that they walk boldly to some for caresses, or avoid others with marked exhibitions of dislike.

Seals gather a distinct meaning of words in their captivity, with a keeper who is regularly in attendance. A change of superintendent leads to sorrowful moanings. By speaking slowly and distinctly to them, their full, intelligent eyes sparkle with evident delight.

Anecdotes, without limitation, are interspersed in works on natural history, illustrative of the capacity of animals for gathering a knowledge of words. Birds certainly converse with one another when preparing for a migratory excursion. They then congregate in multitudes, and apparently deliberate in council. If the chattering means anything, it would seem to relate to the proposed removal to another climate.

If they have no intelligence, and their sagacity is without thought, the force of instinct which compels them to act without the exercise of volition,—how does it happen that such armies of feathered races move with a precision, varying by incidental circumstances, as though they had both a present and future object in view?

Parrots articulate phrases they have heard, without attaching any sense to them. Their power of imitating vocal sounds is surprising. When once familiarized to a routine of expressions, they repeat them, but without reference to their appropriateness to the occasion. They have not a brain for carrying on a train of thought. Mocking-birds are extraordinary imitators. They are extremely jolly and frolicsome over the confusion they create among other birds. Dogs are far superior to most animals, from the circumstance that they retain cerebral impressions, which are recalled for the execution of after acts. A Highland shepherd bids his dog find a missing sheep. Away he scampers in earnest search for it, keeping distinctly in mind what is expected of him. He never can be taught to speak, because he has nothing to say, although he thinks accurately in the line of his special vocation. Having no sphincter muscle to the mouth, labial sounds are impossible.

Monkeys have a vocal apparatus so nearly resembling our own, that it would puzzle a practical anatomist to designate one from the other, if of the same size, when detached from the body. They have vocal cords, well-formed lips, and nerves directed precisely as in the nervous system of the most inveterate talker, yet a monkey has never been heard to pronounce a syllable. While a whip is held over his head, he performs surprising tricks, rides in a circus, goes through manual exercises with a miniature gun, plays a tambourine, collects pennies, etc., but, with all these accomplishments, he never voluntarily practises them when left alone. Neither does he attempt speaking. Their chattering is always the same, whether expressing pleasure or pain.

There are remarkable accounts of dogs, which fully demonstrate their reasoning powers to an extent bordering on the marvellous. Seeing and hearing are special senses, which are exceedingly developed in them, and through which their knowledge is very accurate. A law of limitation puts a stop to mental progress with animals, as it does in respect to their growth. It is quite impossible to educate beyond a certain point, because the instruments for carrying on the operation of thinking are insufficient, either in their number, weight, or structure.

By associating with persons of mild manners, who pet and praise them with kind expressions, some dogs make extraordinary advances. Their canine exploits may be carried so far as to excite both admiration and surprise. Such performers exert themselves under the stimulus of praise, or unmistakably exhibit dejection and mortification when roughly reprimanded.

Some years ago a noble Newfoundland dog, owned by the city of Boston, was kept at the quarantine ground, Rainsford Island, in the capacity of a general watchman. Tiger lived to about the age of twenty years. In that long life-lease, he had acquired a very correct estimate of men and manners. He was deferential to masters of vessels and well-dressed passengers when they came on shore, but if sailors left their boat to wander over the premises, he became demonstrative. They had to run rapidly back to the landing, or feel the effects of Tiger's indignation.

Such was Tiger's ambition to be in good society, that he was invariably on hand to accompany the doctor in his barge when visiting vessels. On account of his prodigious size and imposing aspect, he was usually an object of particular attention, and had excellent bits of meat dropped over the gunwale into the boat, while his master was transacting business in the cabin.

So favorably impressed was that sagacious quadruped with the attentions he received alongside, he occasionally swam back and made a call on his own account. On seeing him paddling around the hull, if a rope were thrown over, he held fast to it with mighty strong jaws, and was thus hauled on board, to the immense gratification of the crew, and no less gratification of the visitor. After being feasted heartily, and wagging his long, bushy tail to those who had bestowed the grub, in a twinkling of an eye he would leap overboard and strike for land.

No efforts were successful in coaxing him back. On several occasions a plot was laid to capture him, but watching an opportunity, as though perfectly understanding the convesration respecting his detention, he gave the sailors the slip by plunging into the surging waves, through which he quickly worked his way to the nearest beach.

In consequence of repeated exposures in aquatic adventures, of which he was exceedingly fond, together with the infirmities incident to age, Tiger suffered severely with earache. The doctor's lady used to make hot poultices, heat bricks, folded in soft cloths, etc., and when ready, she would tell him to take position. He would instantly horizontalize himself under a table, patiently submitting to a satisfactory adjustment of madam's applications.

The transition from pain to perfect relief was never forgotten. Whenever he had a recurrence of the old torment, he regularly whined in an ineffable tone, and shook his head, which was an indication of wanting the kind lady's assistance. It was a curious spectacle to witness his impatience for the brick to heat. Bracing himself in a corner, he would look steadfastly at the red-hot coals till it was taken out, and his mistress announced all ready.

So warmly attached was that sagacious old dog to his benefactress, that he became evidently jealous of attentions shown by persons who did not come up to his standard of respectability. When she stepped from a boat he invariably sprang for the painter, holding it with tenacity till his friend was fairly on dry ground. If one of the bargemen attempted to take the rope away, there was a growl and show of white teeth, that foreshadowed displeasure at interfering with his gallantry. When the lady was quite safely landed, then the painter was dropped, and Mr. Tiger trudged along by her side, as though conscious of having done his duty.

By way of experiment, to ascertain the exact extent of that splendid brute's appreciation of language, on the return of the doctor from the city, madam related to him that during his absence Tiger had not behaved well: he had disobeyed her, or had been over to the hospital and eaten up a patient's dinner.

Such conversation seemed to produce profound sleep, or if he saw that facial expressions were assuming an unfavorable change, he stole away behind a piece of furniture. By tacking ship, however, and praising his fidelity,—recounting his feats and good qualities, the old fellow's organ of approbativeness brought him bolt to his feet. He was delighted with flattery, and overwhelmed his good friend with affectionate demonstrations of regard.

A lady in New York converses with her Newfoundland as with one of her servants. When she says to him, "You may go with John to market," he capers frantically with anticipations of pleasure, because he is quite sure of fine picking among the stalls where he has a host of friends. He goes to the kitchen for a basket, and returns with it for money from his mistress. He then trots off with it through densely crowded streets, safely. His vanity is his weakest point, putting himself at considerable inconvenience for a compliment.

These citations of brute intelligence are digressions, but they belong to that catalogue of evidences which are numerous and convincing, that animals, from canary birds to elephants, in the society of man, unquestionably acquire an elementary knowledge of the true meaning of words.

How absurd then to pretend that a human being, with a great brain, superior in volume to that of all races below him, cannot master more than one language, while donkeys, mules, horses, dogs, elephants, oxen, seals, and even mice and canary birds, gather an elementary knowledge of two or three, without being able to articulate any!

### SYSTEMATIC PERSEVERANCE.

By a very moderate amount of systematic industry, perseveringly continued at leisure moments, any woman may attain a speaking knowledge of one or two languages in addition to her own vernacular.

William Cobbett, alone, without an instructor, became a

critical French scholar. He even wrote an excellent grammar of the language, still in repute. Dr. Franklin acquired French when he was about seventy years old, which shows what may be achieved where there is a will. Such are encouraging examples, showing, beyond question, that it is never too late to learn.

Rev. Mr. Jones, chaplain of the Sailor's Snug Harbor, stated in a public meeting, in New York, that in a company of one hundred and thirty-seven seamen, at his house, they spoke among them thirty-seven languages fluently. Several conversed freely in four; one or two in five; and one, a native of Finland, spoke ten, and wrote seven of them correctly,—one being Latin.

Du Chaillu, the African traveller, stated before the Geographical Society of New York, giving an account of his wandering in Denmark, Sweden, Norway, and Finland, that there is not a girl in those countries sixteen years old, even in the remotest cabins of Lapland, who could not converse readily in English, French, and German, besides their own severely harsh dialect.

That statement made a deep impression on a highly cultivated audience. The whole mystery of such proficiency was explained. Every school, from the primary to the highest, in those countries, is by law obliged to teach children three necessary languages, the government having the good sense to appreciate the value of the principal languages of civilized countries, the knowledge of which qualifies the people to transact business with the ruling nations of the earth.

There is no labored effort on the part of the children, no extraordinary exertion by the teachers, to bring about such proud results. It is simply a gradual process like any other study deemed useful for youth. Really, no child is conscious of any particular effort, but, as a matter of course, they insensibly become accomplished linguists.

Just such a system should be pursued in all common schools in this country. It would be attended with no more expense than at present, with arithmetic, grammar, and other elementary branches. Every little girl and boy, beginning with their *A, B, C,* might speak and read French, German, and English in the same time they are acquiring any and all of those things which make up a common school education.

Let this admirable course be adopted, and in twenty years we should outgrow the taunt flung in the faces of American scholars, that we are a *one-tongued people.* •

Law and medical students, theological also; clerks, and indeed others associated with commercial, banking, and various kinds of activities requiring skill, tact, and accomplishments, too, in their pursuits, have no apology for being so universally ignorant of foreign languages.

By subjecting themselves to a few inconveniences, and taking up a residence in families where a language is exclusively spoken which they wish to acquire, medical, law, theological students, and clerks, might soon speak another language without infringing upon business hours or cost of tuition.

The same course would accomplish any young lady, if not in a condition to pursue the other plan proposed, by going to Canada or France.

### Examples of Success.

Let the organs of hearing be educated. That is the all-important beginning.

A New York lady practising upon this system of keeping the ear familiarized with French, Spanish, Italian, German, and Portuguese, has a servant representing each language, for the sake of being obliged to converse with each one in his or her own language. Her house, therefore, is a modern Babel, the mistress being a ready interpreter in many dialects.

A gentleman went to France a short time since with an accomplished daughter of whom he was exceedingly proud. As the vessel was entering the port of Havre, a passenger asked the father of the young lady if he intended to take a courier into service on landing.

"No, sir," he replied. " My daughter has been expensively educated in French. She understands it like a book, and I intend she shall be my interpreter on our travels."

Very soon after this conversation a pilot and revenue officer came on board. Marching up to the American millionaire,—sputtering at a rapid rate,—he quite confounded the old gentleman with his volubility. Turning towards the blooming daughter, " Find out," said he, " what these fellows want."

She was respectfully approached by the new comers, who stated their object in choice French, but to her inexpressible confusion, she could not understand a single word. The father's mortification could not be concealed. He had boasted so much of her acquirements, and the money lavished on her French education in one of the most expensive boarding-schools, both exhibited fallen crests, as most of the passengers were expecting very gratifying assistance from that source.

The young lady had a thorough reading knowledge of French, far superior, no doubt, to those who had unwittingly brought her into such a mortifying dilemma. But her ear had been neglected, as it always is in those fashionable institutions.

Had she been placed in any French family ten months as a mere boarder, without taking a single lesson, and in no way interfering with other essential studies or social relations, she would have spoken French conversationally with ease and fluency.

## AVENUES TO INDUSTRY.

A man who speaks two languages, says a proverb, is equal to two men, and a woman who can do it is equal to half a dozen.

New routes to useful pursuits are laid open by the aid of a second language. It is an extra key for unlocking a cabinet of treasures.

Young persons should be ambitious to possess that advantage. Population is rapidly increasing, consequently the strife for place and position is becoming more active. Without a speaking acquaintance with at least one more language in addition to her own, a young lady is not equal to the responsibilities of positions she might desire to occupy.

Telegraphic interests in the future will require linguists, and so will mercantile houses, banks, and insurance offices, far beyond what may have been anticipated. Such operators will be in request, so extended are the enterprises of nations since the utilization of steam and electricity.

Young women would be admirable at the wires, and a hope is entertained they may have almost a monopoly of telegraph stations. Therefore, let them seasonably qualify themselves for those useful, appropriate, and remunerative services.

Drive the pale, thin, feminine-looking clerks out of easy-chairs in banks, insurance offices, treasuries, public bureaus, where honesty and faithfulness are the first requisite qualification, to cultivate the soil. It would be doing them a personal kindness. In becoming strong, hardy, brave, and enterprising in the field, food would be cheaper, and the race improve physically, morally, and mentally.

# CHAPTER XXXIII.

## WOMEN IN THE PROFESSIONS.

Not Forcible Public Speakers before Large Audiences—Physical Reason—
Make Good Professors—Female Physicians a Success—Admirable Artists
—Approved Teachers of Various Branches of Education—Should be
Encouraged.

SOME women, unfortunately for themselves, assume unnatural positions. In a pulpit they appear out of place. They
may become learned theologians, write with fervor, but in
standing before an audience, however animated by zeal, or
eminent in qualifications, their vocal powers are not equal to
such occasions.

The larynx is smaller than in men; therefore there is a
physical inability for giving strength to the voice required for
being distinctly heard in large halls or churches, the *timbre* not
being of the quality for ringing through great assemblies.

There is a vast difference between singing and speaking.
The first is appreciable as a musical tone in them, heard distinctly and widely; but when they attempt giving sonorous
weight to the voice, in the manner of commanding orators, the
failure is apparent.

The cartilages of the vocal box remain flexible in females
through life. In men, on the contrary, when they arrive at
puberty, they become bony, and the voice changes from the *vox
rauca* of a boy to a manly *timbre*.

By that organic alteration in the plates of the larynx, there is a vibratory impulse given to the vocal current, set in motion by the vocal cords, more intense than before. No such change takes place in the female larynx.

Again, the nasal cavities, the frontal and maxillary sinuses are far more developed in men than women. They are to the voice what the body of a bass viol is to the strings—as proven by an inflammation which closes them. A vulgar explanation of an alteration of voice is imputed to speaking through the nose, in case of a severe cold, whereas a true state of the case is, that they do not have the assistance of the nose in giving volume and distinctness to articulate sounds.

## ADAM'S APPLE.

Within the protuberance in front of the throat, midway between the chin and root of the neck, is a triangular box, in which ribbon-like cords are stretched from wall to wall, that vibrate by the rush of air, inhaled or expired, passing over their tense edges.

Before puberty the voice of boys is like that of females. They are employed in church choirs, while thus stationary in their vocal apparatus. On emerging from that state into perfect manhood, a change of voice announces what has taken place. The female voice, however, remains always the same, since an evolution from girlhood to womanhood brings no parallel alteration in the larynx or nasal cavities. That organ neither enlarges nor ossifies.

Neither sinuses or nasal cavities are ever as large in females as in men.

A natural conformation, therefore, in the vocal mechanism of the throat disqualifies females for producing a strong, sonorous sound.

There are exceptional cases, in which some women are so masculine in manner, voice, and acts, as to destroy those attributes which make them attractive, and their society sought for their refining influences and loveliness.

At the bar, before juries, in halls of legislation, or, indeed, in large bodies, they could not compete with the deep, loud-sounding voice of a man in the meridian of his muscular power.

By emasculation before puberty, the larynx remains stationary. Its cartilaginous walls are ever after flexible as in females. Eunuchs, therefore, are employed in harems of the East as female guardians ; in choirs also, as singers, where women are not admissible. If emasculation is not performed till after puberty, the system being developed to its maximum of completeness, the voice remains at the *timbre* it had when it altered from the *vox rauca*.

### PROFESSORIAL DUTIES.

As professors of departments in public institutions, colleges, seminaries of any order, where science or literature is taught, no great lung-force being required, females would be abundantly able to sustain such honorable positions. Demonstrations and illustrations would be within their scope. They could discharge all such duties with as much success, eclat, and appropriateness as men, and far more acceptably and clearly than many stupid male professors, who are kept in such institutions through the influence of interested relatives.

Very many institutions of learning in this fair country are languishing. Their perpetual cry is for money. They annoy legislatures for pecuniary assistance. When it is obtained, it does not accomplish the great rescusitating results which were theoretically promised. The real secret of their feebleness is in the faculty, oftener than otherwise, who have neither tact, brains, nor qualifications for the chairs into which they were inducted.

Of all professions, however, that in which women succeed best, is in the practice of medicine. They have made the discovery themselves, that they possess aptitude for managing the sick. The public, too, in this government, and in many of the most polished and advanced governments of Europe, accept the proposition that they make excellent and eminently successful physicians.

Medical colleges have been chartered, in all directions, for the special purpose of qualifying them, scientifically, to take upon themselves the responsibilities of that important profession.

A standing army of medical men have opposed the movement. They have thrown every imaginable obstacle in the way. Not only have they refused to admit them as pupils into schools of medicine, but they have denounced and ridiculed those who have expressed sympathy for them in their desire to be medically educated.

That old saying, "that the blood of the martyr is the seed of the church," is particularly applicable in regard to female physicians. Intensified opposition has created college after college expressly for their benefit, and more will be chartered and they will multiply. Some institutions abroad have been compelled by mandates of rulers to open their doors to them, in direct opposition to the remonstrances of medical practitioners.

Probably there will be as many female medical students in the United States as male students, within the next fifty years.

## Female Physicians.

Communities have to be educated for the reception of whatever is useful. It is extremely difficult to convince the public that improvements are not innovations. The idea of a female physician was a novelty at first, and so strange too, as none but men were practitioners of medicine, it looked like overturning the constitutions of society when women were feeling pulses.

But we become accustomed to revolutions. The novelty wore off, and next it was ascertained that their manner of intercourse with invalids, the delicacy of their approach, the carefulness with which investigations were conducted, the accuracy of their analysis of symptoms, and their judgment in the administration of remedies, inspired confidence.

At once their own sex gave them a preference over many rough, burly, indifferent practitioners, whose attainments were far below the qualifications of those female physicians who have been thoroughly instructed in well conducted medical institutions.

Emigrant ships would immensely improve the condition of steerage passengers, by having a female physician permanently attached to the vessel. Female passengers need one of their own sex, qualified to prescribe for them and their children, and to give them council. It would insure order, neatness, morality, and better health in those crowded collections of men and women across the ocean, were this suggestion accepted. It is an advance in propriety and the comfort of poor, neglected occupants of the steerage, that is bound to be inaugurated, and soon too.

That partition-wall has given way, which prevented the advance of enterprise in law, medicine, and theology. Those professions are now open for all who are qualified to sustain themselves in them.

Women, in being admitted to the privileges of medical practitioners, have wisely let surgery alone. Their gentleness would be out of place where living flesh is to be cut with sharp instruments, even when the object is saving life. Blood is not a sight for their eyes. Let them keep within the boundaries which instinct directs, and their professional ministrations will be appreciated.

Physiology explains the gradual unfolding of the organic system, the function of nutrition, the phenomena of locomotion, vision, audition, and the laws of reproduction. Female medical students as fully comprehend difficult problems as young gentlemen. Indeed, they are ordinarily closer applicants, thus laying a broad foundation for pathological success in their intercourse with the sick.

Either from a desire for distinction, hallucination, or an abnormal craving for notoriety, some women exhibit a perverted taste, and a feeble judgment, when they force themselves into positions which excite ridicule or contempt.

Engaged in pursuits within their appropriate sphere, their success is almost certain. Both honor and profit should accrue to them for whatever they do, in the same way that men are compensated for analogous services.

Women make excellent physicians where their advantages for instruction have been full and complete. They are miserable quacks.

In indentifying themselves with deceptions, whether as the seventh daughter of a seventh daughter, a clairvoyant, a spiritual medium, a magnetic prescriber, who, with closed eyes, pretends to see through an opaque body and detect obstructions in glandular ducts, and such like nonsense,—they fall below contempt.

They are admirable artists, but uniformly fail as quack doctresses. Only the partially ignorant presume to practise without qualifications. When . educated, they are above trickery. Honorable professional industry gives no countenance to hobgoblins or mesmeric nonsense,—instrumentalities of the blind for leading the blind.

Natural philosophy, intimately incorporated with the study of medicine, is an antagonist to superstition. Those who formerly could discern ghosts, and were uncompromising believers in the manifestations of disembodied souls in mystic circles, can see nothing after becoming familiar with the principles of general science.

So-called medical mediums are impostors. They do not emanate from accredited medical institutions. They are not to be confounded with those ladies who have been carefully and scientifically instructed by competent teachers in all the intricacies of theoretical and practical medicine.

Many medical gentlemen, standing well in communities, would run a mortifying tilt with many ladies in an examination before an authorized board of censors, whose decision should depend on the accuracy of their answers.

### APPOINTMENT OF MEDICAL TEACHERS.

Were it customary here to elect professors by *concours*, as in France, and young ladies recently graduated were permitted to be competitors, quite a number of stupid occupants of university chairs, obtained through family assistance or the potency of cash, would have to give way to higher attainments.

In the treatment of female maladies, women are the proper professional advisers. It is grossly unjust to assert that they have no comprehensive views or therapeutic knowledge beyond making water gruel and flaxseed poultices.

In lecture-rooms, wherever they have been admitted, their progress has invariably been equal in all respects to that of young men. The latter dissipate more or less,—smoke, drink to their detriment, ramble to places of amusement, or where there is excitement at evening. Female medical students are not guilty of any such sins, if sins they are. They economize time in thought and study. Better still, they neither chew, smoke tobacco, or stultify themselves with the curse of the United States,—whiskey.

Operative surgery is not their forte. Exact familiarity with the intricacies of surgical anatomy, however, is one of the studies in which female medical students sometimes excel. They pursue anatomy with earnestness, so that they occasionally become experts.

If they have not the requisite firmness or coolness for cutting down into a region of vessels and nerves, nor a strength of arm for reducing luxations, they are quite well qualified to determine the extent, gravity, and probable extent of injuries.

As oculists and aurists, they might achieve great distinction. Ophthalmic operations are neither bloody, very painful, or attended with hazard to life. With their natural delicacy of touch and thorough acquaintance with the structure of the eye, they might cure deafness and extract cataracts just as skilfully as operators of the other gender. The first competent female oculist who commences under favorable auspices, could not fail of success,—if the accumulation of a fortune were the evidence of it.

This suggestion is for their consideration. They must expect to encounter opposition; be misrepresented and abused, because it will be a novel interference with the imagined prerogatives of those specialists. It is always far more remunerative than ordinary general practice.

Why could not women become expert dentists also?

Their tact in watchwork, the manufacture of rich jewellery, penmaking, and some other artistic employments with which they are identified, besides modelling, designing, painting, and engraving, in each and all of which they succeed admirably, insure equal success in the practice of dentistry. Filling carious teeth, inserting artificial ones, taking casts of gums, restoring cleft palates by the insertion of metallic plates, etc., are all within the sphere of their genius.

Certainly women draw, etch, color, conduct photographic and lithographic establishments; and what is to prevent them from extending the area of honest enterprise? In each and all of those callings they could earn, legitimately, quite as much as men, and what is to prevent them from being equally well compensated?

### MENTAL ACTIVITY.

Great undertakings are not accomplished by main strength. Brain force is that specific power wielded by orators, influential divines, brilliant commanders, revolutionizing writers, disturbing politicians, great property-gatherers, bold projectors, and by all those men and women who leave ineffaceable memorials of their existence in the archives of history.

Neither legislative enactments, denunciations from the pulpit, the bitterness of reviewers, the decisions of unjust judges, or the giant strength of money, can stay the march of genius. It is stronger than all mechanical powers combined. Genius is not boisterous or presumptuous. It is a quiet faculty. Pretenders are both positive and superficial. The records of history and the experience of mankind prove that women, in capacity, originality, diligence, thoroughness, skill, and intellectual acumen, are capable of accomplishing in art, science,

and literature, whatever men do under precisely analogous circumstances. They are therefore entitled to the privilege of
embarking on the sea of enterprise.

Men who are rated vastly beyond their merits, harnessed
in petticoats, laced in stays, half-clad in gossamer garments,
corded round the waist, and dieted on dry toast and tea,
restrained by arbitrary custom to the house eleven hours in
twelve, breathing impure air, instead of refreshing, vitalized
currents out of door, mounted on high heels, and every hair on
their heads put upon a stretch, and held back by iron pins and
combs,—suddenly called, would they appear to any better advantage than their mothers, wives, and sisters?

## PUBLIC OPINION A RESTRAINT.

Women are cruelly hampered and restrained by the fear of
what may be said of them, so that physically and intellectually,
they appear to disadvantage.

There are some generous enough to admit that women have
natural rights of which they have been defrauded by lawmakers. With the progress of liberal sentiments, a gratifying
feature of modern civilization, concessions are gradually being
made to them. A restoration of rights and privileges must be
made.

Theoretically—and it is a legal fiction—a man and wife are
one; "but the husband is the *one*," said a female orator on a
notable occasion.

# CHAPTER XXXIV.

## MARRIAGE.

THE great event in a woman's life is marriage. They reckon from the epoch of their marriage as a point of departure. It is the first milestone on the highway of domestic relations which outranks and overtops all other circumstances in their earthly pilgrimage.

They begin to think of it early, without having any very definite views of the responsibility that belongs to that solemn connection.

Universal attention is given to the subject in all countries, yet only a few of the many marry precisely to their liking.

Were it possible to obtain a true and exact knowledge of the amount of domestic happiness appertaining to that state, wedlock would make some strange revelations.

Very excellent ladies, model women in their matrimonial relations, are often wearing a mask to conceal a cancer gnawing at their heart.

They are compelled to be hypocrites to the end, because respectability is everything. To assume the appearance of happiness, prevents the mortifying comments of those in whose estimation it is an object to stand well.

Merchants, bankers, and, indeed, most men in active business who give employment to young men, keep them at an unwarrantable distance. The civility of inviting them to their own houses, and giving them the acquaintance of their pleasant families, rarely occurs, however much their clerks may have merited their esteem. Some become dissipated from having no respectable places to visit,—none to give them an encouraging recognition.

How many such neglected counting-room drudges become the leading men of the day, eventually taking a flight entirely beyond the narrow circle from which their patrons excluded them! Splendid husbands might have been discovered in such neglected worth, by attachments formed between lovely young ladies and poor but deserving young men. The policy of allowing those with nothing for a capital but unsullied honor and enterprise, to address a rich man's daughters, by no means has the approval of a managing mother. Her ambition is to engineer her angels into favor with those reputed to be worth the most.

Neither heart nor principle is involved in the speculation, as matrimonial adventures are now conducted on both sides of the Atlantic. Women are notoriously bought or sold to the highest bidder. Love is not in the bargain. The purchaser obtains a fool with her dot, and she a rake who wishes her under ground after getting control of the funds. Those are the matches ending in divorce.

Were young ladies satisfied with the attentions of virtuous, unpretending young men, whose only fault is their poverty, what gems they would often secure! General Washington offered himself to a lady to whom he was devotedly attached, but had the mortification of being rejected by the haughty heiress, because he was only a major without property. She afterwards stood at a window, in the city of Balti-

more, as that spurned lover passed through the street, lined
on either side by immense multitudes with uncovered heads,
—President of the United States of America, the saviour
of his country, whose name and fame will live till time shall
be no more.  She swooned, and was removed from the apartment.

In marital relations, women carry the heaviest end of the
beam.  They are too much burdened in the middle walks of
life with cares, and consequently they suffer more than men in
family responsibilities, especially when uncongenially united.


## EXCUSES OF BAD HUSBANDS.

Husbands absent themselves from disagreeable homes on a
plea of business, when an apology is necessary.  Frivolous pretences are always to be found for absence without exciting particular remark that might essentially affect their moral standing
in society, were the exact facts of the case known.

Their wives, however badly treated or neglected, cannot
flee so readily from the presence of one who abuses them,
without raising a whirlwind of ungenerous surmises injurious
to their reputation.

No true woman likes to face a tornado of scandal.  Men
and women must associate.  It was so designed from the beginning.  Monastic institutions, which interdict matrimony, are at
war with nature.  It is unnatural and opposed to a fundamental
law of life.

A society which forbids the association of males and females on a basis contemplated in this proposition, cannot maintain such a system of discipline without exercising a vigilance
perfectly despotic.

## SHAKER CELIBACY.

The Shakers have probably carried the experiment as far as, any anomalous religious sect, and as successfully too, so far as complete non-intercourse is essential in their creed. They make but few converts, and the sect would soon be extinct were it not for the children they gather among sinners.

Their organization is recruited from sources they absolutely condemn as sinful.

With their wealth, their beautifully cultivated farms, model gardens, well-finished brooms, medicinal herbs, carefully packed seeds, apple-sauce, and some other manufactures, their societies may be held together through one or two more generations. Thus all religious monastic associations are sustained—in opposition to a law of nature.

There must be, inevitably, a last day in their calendar. A dissolution will come, because they are opposed to a law of God, on which the perpetuity of races depends. It can neither be modified nor repealed by human effort. Let the Shaker doctrines be fully sustained, and the beautiful earth in two hundred years would not have one human being on its surface.

Institutions antagonistic to laws governing our physical organization, cannot be sustained. There may be a temporary show of resistance, and a pride in pretending that extraordinary exaltation of mind is attained by subordinating all emotions, passions, and instincts to the empire of reason, but nature triumphs at last.

Religious enthusiasts are prone to announce theories which they proclaim to be decidedly gratifying to the Creator. Can it be a pleasure to that ever-living Power, that has fashioned things as they are and as they will continue to be, to have intellectual beings in perpetual warfare against instinct, with which he has endowed us? Death alone can give them their quietus!

Still, with a knowledge of the physiological endowments of nature, they make unrelaxing efforts, and bigots run mad with pent-up wrath, because they cannot rule supremely and force their dogmas and crotchets down the throats of unbelievers.

## MARRIAGE.

A majority of mankind, wherever located, savage, barbarous, or enlightened by education, act in strict accordance with natural laws, conducive to health, morals, and happiness.

It is not necessary to discuss the subject to prove the truth of this proposition.

Thus it has been since Adam resided in the garden of Eden, and so it will be while men and women are in existence.

Matrimony is an ancient institution, but the misery of being mis-matched is a condition of wretchedness, which sentimental reformers will find it hard to remedy. If a couple are joined in wedlock, and subsequently discover that they are mismated, it is a tedious process to relieve themselves of the misery of that relation.

In countries making no pretensions to civilization, when parties discover they are not congenial to each other, they simply separate. That is a relief which the civil law, and certainly ecclesiastical law, very tardily and reluctantly permits.

Were it possible for contracting parties to understand each other before marriage, in regard to temperament, disposition, moral feelings, and tendencies of character, it would be a blessing to both, since they could avoid many of those forms of unhappiness that lead to dissatisfaction, hate, and, lastly, revenge, which occasionally closes the drama of married life, before the real purpose and responsibilities of the compact are understood.

Some women imagine their lovers are to continue picking up their gloves from the altar to three score and ten. Silly men appear surprised that the angels they have caged wear shoes, and actually possess stomachs. Of course, where there is neither judgment nor common-sense for guidance, there is no binding principle.

Assuming that men and women were designed to live together, it is a problem with divines and legislators how to regulate marriage so as to secure equal rights, without caring a fig about the domestic happiness of those entering upon that solemn contract. The law looks after property, and regulates the disposition of it in every contingency, growing out of the discontent or separation of those who have been legally joined.

No combination of circumstances produces such real felicity as the relation of husband and wife, when congenially united. If not cordially associated, then it is an intolerable bondage, hard to bear.

Ecclesiastical laws contemplate a secure binding, leaving the parties without escape from miseries which may, and certainly do, follow inharmonious marriages.

When united, as it has repeatedly happened in this country, as the parties supposed, in jest, on a festive occasion, and it was subsequently discovered the ceremony had actually been performed by a magistrate, unknown to them, in that official character, it has been held in law that they were husband and wife, although entirely contrary to the wishes or expectation of both sufferers.

Such a connection, it would seem, in equity and reason, ought not to be obligatory. But the civil, and, perhaps, ecclesiastical tribunals concur in holding the parties to all the obligations that bind consenting, loving couples.

Such marriages cannot be dissolved without wading through tedious and expensive legal processes, and a free expenditure of money, which last power carries more weight with it than the eternal principles of justice.

Communities are agitated by occasional accidental marriages. Sometimès a deep plot is laid, and one of the parties is unsuspectingly duped into the trap matrimonial by an irrepressible lover or fortune-hunter.

Courts are invoked, and legislative bodies implored for special acts for emancipation from wretchedness that can only terminate with death, if no relief is afforded at the fountain from whence the law had its origin.

Nothing can be accomplished by the unhappy poor in that dilemma, but what they do for themselves. Women commit suicide, and men run away beyond the knowledge of those who may have known them. It is useless for those without funds to pray tribunals to break the chains that hold them in uncongenial wedlock, and let them go free again, even when the petition is a mutual prayer of the aggrieved sufferers.

### Purchase of Relief.

Paving the way to justice with dollars, is the modern method of making a quick passage over a rough road. Money is omnipotent with magistrates, who care more for mammon than the approval of conscience.

There is a cry for a modification of certain laws for the special benefit of the unhappy in marriage, to arrest the progress of demoralization in the land.

It is asserted by Oriental travellers that there is far more domestic felicity in those far off, unchristian countries, where wives are purchased, and even among savages, whom instinct, and not sentiment, guides in the choice of a wife, than with us, where law binds, but reluctantly unbinds, the tangled web of infelicities, which sometimes characterize matrimony.

A grave question with moralists is this, viz.: Has any government positive, unquestionable authority for imposing obligations upon men and women, that it would be a violation of a divine law to abrogate, if they failed to secure the purposes for which they were enacted?

There is such manifest dissatisfaction all over the United States, with legislative action respecting matrimonial affairs, that some new legal principle is urgently demanded to meet the emergency.

As it is, divorces are as common as revolutions among dissatisfied politicians. Even the descendants of the Puritans, priding themselves on their law-abiding character for propriety, have become restless. Nowhere is there more wretchedness in matrimonial bonds than in the New England States,—largely tracing their origin to the voyagers of the Mayflower.

### HELPMATE.

Not unfrequently the press represents that women are oftener to blame than men, in squabbles for emancipation. The cry comes from all points that female education, as now conducted, quite ignores those homely notions once in repute, that a wife should be a helpmate.

Women have witching powers of fascination for leading silly, as well as well-balanced intellectual men wherever they choose. But the contrary creatures themselves cannot be driven an inch.

In this fact is discoverable the origin of many family troubles, culminating in ineffable misery, which nothing short of divorce can assuage.

Climate exerts a baneful influence on some temperaments at the North. Nearer the tropics, divorces are less frequent, even among those who have no educational advantages.

The clergy, claiming to interpret the Divine Will, unwilling to relinquish their hold upon the masses, are accused of keeping old theories and old customs alive too long.

Law or no law, human or divine, when a couple discover an incompatibility for each other, they generally act independently of legislative requirements,—braving the denunciations of the pulpit.

Where obstacles are interposed which cannot legally be removed for the accommodation of one of the parties, enormous crimes are often perpetrated under the idea of regaining lost liberty.

Cruelties, suicide, and murder are the bad consequences of compulsory laws, obliging those to drag out life in riveted wretchedness who desire separation.

## A QUESTION REGARDING DIVORCE.

But divorces ought not to be granted on a trifling pretext. When a man and wife declare their determination not to reside together, for reasons best known to themselves, cogent and right in their own deliberate estimation, what is gained for public morals by keeping their shackles riveted for ever?

Neither society, religion, nor the State, is benefited by an unrelaxing policy which would see both ruined for earth, and unfitted for heaven, in the agony of their uncongenial condition.

While we are unflinching advocates for marriage, based on affection, it seems cruel to open no avenue for escape under circumstances which plead for sympathy and relief.

It is useless to attempt the development of love or personal respect by law. A physiological argument, that celibacy is unfavorable to longevity, never hurried any one into matrimony on the ground that it was solely for the purpose of saving life.

## MORAL REFLECTIONS.

Men and women unmarried have a weaker hold upon life than the married. Were the institution abolished, public health and public morals would reduce society first to confusion, and then to chaos.

Mankind cannot be sustained in soundness without obeying laws on which the perpetuity of a race depends.

Marriage, then, not only elevates humanity, but also gives us a stronger hold upon life. Single men or single women do not live as many years as the married, all things being equal, nor are they as free from indisposition on an average.

A minute exemplification of this assertion would be too professional, hence illustrations are omitted. Every medical practitioner could verify this statement from the records of his own practice, were it necessary.

In 1869, the following statistical information was chronicled in Illinois, abundantly proving as much as may be required for sustaining an opinion that matters are loosely conducted in one State, if not in all, calculated to rouse the apprehension of moralists in regard to the future condition of a Christian community :—

" Two hundred and seventy-four aspirants for widowhood, out of a total of 454, filed their papers in that court, and 195 discontented husbands appealed to the same tribunal. The whole number of divorce suits commenced in the three courts in 1869, was 723 against 430 the previous year. Four hundred and fifty-four of these were brought by wives, and 269 by husbands.

The 'better half,' it will be seen from this, is the most restive half in the hymeneal coupling by considerable. Of the 454 wives who sought release from the yoke of matrimony, 304 were made happy by liberation, and 150 were remanded back to the galling bondage. The husbands fared somewhat better in proportion, as they generally contrive to do, and 191 out of the total 269 were sent on their way rejoicing to seek new affinities."

Life-insurance companies are gradually gathering in a mass of statistical illustrations confirmatory of these views, some of which are quite new in their physiological bearing; at least, they have never been so plainly and intelligibly demonstrated in a popular form. They are hygienic discoveries, sent forth like pilots to guide those who are authorized to take risks in the issue of policies.

Unfortunately for themselves, women seem to consider maternity a disease, and, therefore, accompanied by a hazard which many are unwilling to incur. Confinements, slight and temporary as they are in ordinary childbirth, are contemplated by insurance offices as sickness, perilling life. Childbirth is not a disease. It is not a condition that should give the slightest apprehension of danger. To become a mother is equivalent to having a longer lease of life. The oldest women are those who have borne many children.

Formerly, those who had had the largest number were most honored. Now, those having the fewest, or none at all, are complimented as fortunate beings.

## HAPPINESS IN CHILDREN.

Children are not a curse, though they are sometimes sources of great solicitude and care. On the contrary, they are blessings, even to those whose means are most restricted, as might be shown were it of consequence to reiterate what is universally admitted to be true in all countries and among all orders of men.

Large families present a strong front, but a childless household is a desolate place before the sands of life have run out.

Children are a national blessing. Mothers of many are the safety of a state. Those without them have contributed nothing to humanity. Who is to rise up to call them blessed?

Examples of extreme longevity have been recorded of females who had never been mothers. Such, however, are exceptions to the general law of feminine life.

If it could be ascertained what the precise condition may have been of those represented to have died in childbed, it is probable it would appear they died in most instances from other causes. During gestation, tuberculous affections of the lungs and scrofulous difficulties that were undermining the constitution, are usually partially suspended, that the new being in its embryotic state in utero may be developed. After its birth, the malady kept in abeyance then resumes its destructive course.

Nature steps in with a helping hand, keeping back the messenger of death till the new candidate for life comes into the world to be sustained independently of the maternal system.

This is a feature of such striking import, that it cannot be viewed in any other way than a special provision for meeting an emergency.

It is a further subject of curious philosophical interest, that medical reminiscences also furnish proofs of perfect restoration

to health, from feeble emaciation, during the months of gestation.   The system had time for recuperating during a suspension of a disease ; and, once gaining an ascendency, the vital
forces were able to maintain the advantage after parturition.


## LIFE EXPECTATION.

By marriage, the expectations of life are enhanced and improved.   In that relation, women live longer than men.

Widows have more vitality than widowers, all other things
being equal, and a majority of them are alive, when widowers
are very considerably reduced in numbers, in any given area of
country.   This difference is due to the better habits of women.
They are more serene, secretive, and less exposed to debilitat-
ing excitement.

Women are more reserved than men, less frequently thrown
into abandoned society, and when their suspicions are roused,
that contaminating influences are approaching, they resist demoralizing attacks far more heroically, besides being character-
istically more consistent and conservative.

Single ladies, especially in New England, are prone to engage in reformatory schemes.   They are prodigiously resolute
in their efforts to compel the world to believe in their pre-con-
ceived standard of right.

Other States furnish a few strong female representative
minds, devotedly working to keep the social elements in commotion, that they may finally settle down in conformity to
their theoretical ideal of political, social, and moral equality on
earth.

Married ladies less frequently exhibit themselves on plat-
forms as agitators.   When they do ascend the rostrum, however, and once put their hands to the plow, there is energy, but

always less powerful than exhibitions of single women, when they are fully persuaded they have a mighty mission to perform.

When a married woman dives into a sea of political strife, or rouses the community by sensational appeals, were the precise motives influencing her truly known, it would be found it was only a safety valve, by which she endeavors to make an escape from some domestic infelicity at home, or to divert the attention of the public from herself to herself. It is not applause that is sought. She is striving to conceal something, the publication of which might plunge her to the depths of unhappiness.

The appearance of ladies in unnatural positions, officiating as political orators, reformers, preaching, figuring as military officers, and similar performances, which their organization, thin dress, education, and habits forbid, in the judgment of mankind, and the promptings of their own feminine instincts,—it is morally certain there is something to be concealed.

On ascending the pulpit or the forum, either to plead causes or expound theology, it may be assumed that those positions are only waste-gates, through which are floated away pent-up nervousness.

### WOMANLY AFFECTIONS.

Yearnings of the heart in women require objects on which to bestow those outgushings of love, which belong to their nature. If disappointed in youth, the fire never goes out, even in advanced age.

It burns with intensity in middle life, but may be modified by new relations, which divert the mind from a perpetual contemplation of unrequited affection, cruel neglects, or slanderous insinuations, which embitter the soul.

"Earth has no rage like love to hatred turn'd,
Or hell a fury like a woman spurned."

In a prize essay on the physical signs of longevity in man, published in 1869 by the Popular Life Insurance Company, it is laid down as a remarkable circumstance, that when the glow and warm blood of youth have cooled with an increase of years, single women are exceedingly prone to embark in some radical scheme or *ism*, quite in contrast with their former tenor of life.

If, says that paper, they embrace religious or political doctrines, quite unthought of, or which, perhaps, if reflected upon carelessly, had made no permanent impression, there is no calculating upon the force of their enthusiasm.

In Europe, an excess of vitality in the sex is exhausted in some other direction.

New-England women find no outlet to their excessive accumulation of mental force giving such immediate relief as facing assemblies of dissatisfied persons like themselves.

Single women, however mentally moved to revolutionize a village or the State, with all the strain brought to bear on a fragile system in the promulgation of the cause they may have espoused, have a stronger hold on life, and a better prospect of old age than single revolutionary men, simply because they neither smoke, chew tobacco, drink to excess, carouse at places of entertainment, or keep very late hours in protracted excitements.

Their regularity in diet, and freedom from common dissipations which disgrace men, are anchors that moor them safely in a sea of social commotion.

While on this subject, it may be of some service to female readers to have the views of discreet medical statisticians on conditions which are inherited, affecting their longevity, drawn from the same source from whence was extracted the paragraph on female reformers.

## Probabilities in Regard to Life.

*First.* Both men and women, born of a parentage remarkable for long life, inherit vitality, and are generally tenacious of life.

They occasionally reach a very advanced period, being rarely the victims of acute epidemic diseases.

*Second.* Children born of parents, one but not both of whom inherited long life, do not equally inherit vitality.

In a considerable number of brothers and sisters thus born, some of them will live to be aged, but not all.

*Third.* Men or women with particularly long bodies, otherwise well developed, and governed by all the circumstances and conditions heretofore noted, give satisfactory physical signs of a long life.

*Fourth.* Married women who have been mothers, if in comfortable circumstances, especially in the country, have the prospect of a longer life than those who have not borne children.

*Fifth.* Widows have not the prospect of so long a life as married women.

*Sixth.* Widowers have not a prospect of so long a life as married men. Married persons, if happily connected, have a prospect of a longer life than if unmarried.

*Seventh.* Unmarried women, in health, easy in their circumstances, and pleasantly conditioned in society, have a prospect of a longer life than unmarried men of the same social standing.

*Eighth.* Unmarried women, dependent upon their own personal efforts, and harassed by anxieties, have not a prospect of long life.

*Ninth.* Excitable, fractious men or women, when married, who are subject to paroxysms of sudden anger, peril their prospects of a long life.

*Tenth.* Both men and women, although in easy circumstances, if of a jealous, irritable disposition, or subject to morose exhibitions of temper, married or unmarried, have not a prospect of long life. Still, a few out of many may sometimes live to be aged.

*Eleventh.* Men or women who have changed their residence from a cold or moderately temperate climate of one continent, to a similar one on another, if comfortable in circumstances, and industrious and correct in their habits, do not have their vitality impaired.

*Twelfth.* Men or women who remove from one continent to another, as from Europe to America, or from America to Europe, if inclined to excesses which impair vital force, may die prematurely.

### EXCESS OF FEMALE POPULATION.

Females in the New-England States already outnumber the male population at particular points; and there is a social cause operating that will give a female majority in all of them within a few years.

An excess of females in nearly all the large cities of the Atlantic coast, from Maine to Washington, is an unfortunate circumstance for the prosperity of the nation, as it is for humanity.

It is impossible for them all to have husbands, simply because there are not men enough, numerically, to meet the case, provided, in all other respects, the way were clear for honorable marriage.

A discrepancy is an argument for providing for an increasing army of females proper and remunerative employments. It must be done, or fearful consequences, poverty, destitution, demoralization, crime, and, indeed, a deplorable moral desolation, will certainly ensue.

If they could be induced to cast their bread upon the waters of hope, in the fruitful regions of the West, where men are vastly more numerous than women, their protection would be complete, and they might safely calculate upon that measure of security, happiness, and ultimate independence, which flows from virtuous and well-directed efforts.

Here is a statistical synopsis of the population of the globe, with a classification.*

---

* There are on the globe 1,288,000,000 souls, of which

360,000,000 are of the Caucasian race.

552,000,000 are of the Mongol race.

190,000,000 are of the Ethiopian race.

176,000,000 are of the Malay race.

1,000,000 are of the Indo-American race.

There are 3,642 languages spoken, and 1,000 different religions.

The yearly mortality of the globe is 33,333,333 persons. This is at the rate of 91,554 per day, 3,730 per hour, 62 per minute. So each pulsation of the heart marks the decease of some human creature.

The average of human life is 33 years.

One-fourth of the population dies at or before the age of seven years.

One-half at or before seventeen years.

Among 10,000 persons one arrives at the age of 100 years; one in 500 attains the age of 90; and one in 100 lives to the age of 60.

Married men live longer than single ones.

In 1,000 persons 95 marry, and more marriages occur in June and December than in any other month of the year.

One-eighth of the whole population is military.

Professions exercise a great influence on longevity. In 1,000 individuals who arrive at the age of 70 years, 43 are priests, orators, or public speakers, 30 are agriculturists, 33 are workmen, 32 are soldiers or military employés, 29 advocates or engineers, 27 professors, and 24 doctors.

It is evident that there is no time to lose, if there is a ray of ambition to turn life to the best account.

There is not only a perpetual yearning for something more than we have, but a strife also for positions that promise, either truly or theoretically, to facilitate the acquisition of that in which much happiness is imagined to exist.

---

Those who devote their lives to the prolongation of others die the soonest.

There are 336,000,000 Christians.

There are 5,000,000 Israelites.

There are 60,000,000 Asiatic religionists.

There are 190,000,000 Mahometans.

There are 300,000,000 Pagans.

In the Christian Churches—

170,000,000 profess the Roman Catholic.

75,000,000 profess the Greek faith.

80,000,000 profess the Protestant.

# CHAPTER XXXV.

## Their Dangers in Marriage.

Laws of Descent—Evolution—Marriage of Whites with Blacks—Mental and Physical Deterioration—Manly Perfection—Inherited Virtues—Selections, etc.

How far and to what extent we are accountable for what, to our short-sightedness, seems quite beyond control, is a question to be pondered upon by those who assume to be wise where others are in doubt.

Our existence is forced upon us. It is a destiny, and, therefore, no way within the sphere of our volitions.

Were we consciously alive before being united with perishable humanity, and it were optional with us to change relations, and become associated with a body subjected to the vicissitudes which are inseparable from existence on earth, how many would probably hazard the enterprise?

Nothing in the divine economy is more marvellous than the succession of animals and plants.

Wonderfully ingenious contrivances are invented, which perform operations so complicated and extraordinary, that an unsuspecting observer would be ready to admit that the movements indicated a spirit of intelligence. Such may be the combination of wheels, springs, and weights, as to appear like the phenomena of life. And yet, life surpasses the comprehension of the profoundest investigators, and the most learned in science. Ingeniously devised as machines may be, none of them keep themselves in repair, or reproduce similar machines when the old ones are broken or worn out.

Nature, superior and supreme, does both. One generation succeeds another, supplying the world with new and vigorous laborers for uninterrupted progress. The fountain from whence flows a river of life is exhaustless. Though man dies, and individuals are forgotten in the revolutions of time, yet while the globe moves in its orbit, men will be in existence to superintend the domain to which they belong.

With the progress of discovery, we have had glimpses of wrecks of ancient cities, and examined skeletons of monster animals, that once had exclusive possession of this fair country, at a period so vastly remote, that neither chronologists nor geologists agree upon the number of centuries those osseous remains have been hermetically sealed up in rocks, or concealed in the bowels of the earth.

Marine shells on the summits of the highest mountains, raised to their lofty elevation by upheaval forces from the depths of primitive seas, testify to mighty revolutions in the physical aspect of the land and sea.

### Evolution.

A query has been advanced as to whether the lineal descendants of any progenitor in families now recognized as representatives of ancient types, bear a resemblance to those from whence their existence was derived.

Learned inquirers contend that there has been a gradual evolution going on from the very creation of each and every race now in existence, and, therefore, the last in the series must be entirely different in structure, and, consequently, has modified tendencies, instincts, and propensities.

This is a new doctrine with an increasing school of disciples.

From the simplest forms, according to the new theory, complicated structures and exterior forms far more perfect have been developed, and, therefore, better calculated for sustaining higher relations than those from which they originally sprang, far back in the realms of chaos. Germs could not have possessed either volition or locomotive force.

Assuming that man was at first a granule, a mere speck, a germ floating in a fathomless, illimitable ocean of space, in which was embodied an inherent vitalization, always exerting itself by unconscious efforts to push out further, and to become larger, stronger, and, perhaps, have organs of prehension, it is quite as difficult to manage the problem of a first commencement of the spark of life, as to account for the manifestations of intellect.

How long men have walked on two feet or had a brain capable of reasoning, eludes the prying industry of paleontologists. Some are becoming bold in their determinations to ignore the Mosaic cosmogony. They pretend to believe that man has been on this earth far longer than the sacred historian represents, if a true interpretation has been rendered of the inspired narrative.

If the mastodon, and the great saurian reptiles almost one hundred feet in length, were extinct ten thousand years ago, some have the presumption to assert that man was here with them.

A few arrow-heads, found sticking fast in a skull of a gigantic monster that, theoretically, has been dead ten thousand years, is brought in support of the proposition. They assume it as almost conclusive evidence that men of those times were hunters, and that flint-armed arrows were fabricated by them for killing game. That was the stone age.

It might be asserted with equal propriety, that those animals have not been extinct one thousand years.

Let all speculations of that kind pass, since our geological acquirements are not so firmly fixed but they may undergo many modifications in the progress of further discoveries.

Theories are easily constructed and unceremoniously abandoned without loss. We have penetrated but a little way into the crust of the earth, where strange things will come to the surface to astonish naturalists at some remote future.

## LAW OF DESCENT.

Transmitting to a new being some anomaly recognized as an anomaly, because of a striking deviation from the type to which it belongs, must be received as accidental, and not in accordance with the laws of descent.

Were cross-eyed parents invariably to have cross-eyed children,—hair-lipped sons or blind people, the offspring of persons thus unfortunate,—it would give some coloring to the speculations of those who insist that there were five Adams, progenitors of the five known races of men. There are indeed five distinct races. There are certain peculiar characteristics by which they are readily distinguished from one another.

One is yellow, a second black, a third red, a fourth white, and a fifth something else. These are perplexing facts; but on the supposition that climate has produced alterations which have become permanent, is the way the subject is most readily disposed of by those who give it the least consideration.

According to the record of Genesis, the first man, Adam, was created about six thousand years ago. That belief is sustained by researches not to be lightly questioned.

Here we approach stumbling-blocks, that derange many

finely-drawn arguments, not strengthened by science. Revelation is one thing, and the laws of nature something entirely different.

About two thousand years after the creation of Adam, some of his lineal posterity were singularly altered, in the color of their skin, if it is assumed he was a white man. If the Caucasian is a type of our Eden ancestors, strange changes have taken place in the form of the face of the representative races of men now in existence. Monuments are still standing, four thousand years old, inscribed with characters which record, probably, remarkable events. Enough of some of them have been deciphered to ascertain their immense antiquity, reaching within two thousand years of Adam's lifetime. And on many of them are sculptured facial outlines, profiles, and human faces, that show men looked then just as they do in eighteen hundred and seventy-three. Five distinct races of human beings unquestionably existed then, that is, four thousand years ago.

The negro features on those monumental guide-posts into the obscurities of the past, were precisely what they are in Africa to-day. The protruding jaws, thick lips, and crispy, woolly covering of the head, were then as they are now.

How was a change from a Caucasian type, if that was the original facial form, color, and expression, brought about in two thousand years, and from that period, resulting in permanent alterations, that arrange mankind in five distinctly marked varieties of men?

No essential physical, and, possibly, no moral tendencies or changes have occurred in four thousand years, since that grand revolutionary alteration of the primitive outline form. Nor is there any reason for expecting further modifications in four thousand years to come.

Monster children rarely live more than a few hours from

birth. Those born defective in limbs, or with peculiar markings, misplaced viscera, harelip, etc., in becoming parents, do
not transmit their defects to their offspring. They are as fair
and as beautifully proportioned as the children of symmetrical
parentage. The exceptions to that law are referred to in another chapter, as anomalies.

Chickens are hatched with two heads, four legs; or a boy is
born with only one arm. But they do not propagate those deviations from a normal pattern, which is characteristic of a species. They have no descendants like themselves.

That would eventuate in confusion. The fair world we inhabit, were there no fixed laws respecting definite forms,
would soon team with hideous monsters, widely differing from
one another, both externally and internally.

Order being an unchangeable law, any deviations from a
primeval standard, if varying at all, must be very gradually effected, requiring the revolution of centuries upon centuries.
There is no sudden or violent departure.

A mule rarely, if ever, propagates. While some naturalists
claim it possible, others are strenuous in the opinion that
it would be impossible, inasmuch as it would be a violation of a
fundamental law of nature, perpetually in force to preserve
races, and to prevent anomalous admixtures of blood, that
would lead to an animal chaos.

Sterile as mules are, they are influenced by instincts and
propensities, peculiar to the two distinct stocks from which
they sprang. There is a compensation for their anomalous condition,—their longevity exceeding both horse and ass. The latter have the pleasure of rearing others to take their places,
which the mule cannot have, as the maternal parent has in
nursing and protecting her long-eared colt, singularly unlike
herself in exterior appearance.

An ass is old and quite stupid at twenty, although his lon-gevity is beyond that of the horse.

A mule, not abused, is hale, strong, and serviceable, at fifty. They have reached eighty years. Though faring poorly, and usually treated with severity, he has a compensation in immunity from ordinary equine maladies.

A mixture of blood among different races of men neither promotes health, strength, nor longevity.

Some singular phenomena present themselves in the amalgamation of Asiatics with Africans or American Indians, which have a bearing on moral questions, that must necessarily be met by those who are earnest for the improvement of humanity. But it is a topic to be approached with extreme caution, to avoid shocking the sensibilities of modern political philanthropists, who discover no difference in the intellectual calibre of the white, black, red, or yellow man.

That a soul may be encased in different-colored envelopes, according to climatic influence, is a doctrine taught by radical social reformers, without affecting its powers.

We shall not discuss that subject, which has invariably been productive of more vindictive feeling than sound philosophy, whenever brought forward.

One of the evils attending a practical illustration of the doctrine, that it is perfectly right and proper to amalgamate races, and mix those of different color and facial expression, according to the fancy of unreflecting parties, is a positive certainty of deterioration, and the final disappearance of those whose origin is thus commenced.

Is it no violation of a natural law of which each and every one, however low in the scale of intelligence, has an instinctive appreciation, for whites and blacks to intermarry?

Is it not wrong to rear families of all intermediate shades, whose parti-colored appearance does not meet the approval of either party?

Their children are born to a conscious feeling of degradation.

## MARRIAGE OF WHITES AND BLACKS.

All men are born free and equal in the sight of God, and, in the language of political orators, have a right to pursue the way that leads to happiness.

But where is the most devoted friend of the most oppressed and maltreated of all races, the negro, who would not manifest a repugnance to the union of his accomplished daughters with black husbands, however unexceptionable in manners, culture, or character?

To pretend that no such sentiment as an instinctive objection ought to operate against it, would give the lie to one of the strongest dictates of nature and conscience.

For the sake of appearing consistent in the estimation of those who might comment to the disadvantage of those promulgating the doctrine, that color should not be objectionable in forming marriage-ties, teachers of such abominable sentiments may successfully conceal their true feelings; but they live hypocrites, self-condemned.

We cannot go counter to the established laws of nature and morality, without having a conscious prompting of the wrong we have been doing.

In a first remove from the mixed parentage of black and white, the children are not all of the same tint. Among a group of six, for example, one may be black, with protruding lips and short woolly hair; another will have a retreating forehead and lighter complexion. Neither the features of the father or the mother are distinctly reproduced in either, while all of them are marked deviations in form, stature, color, and, perhaps, mental calibre, from the parents.

## MENTAL AND PHYSICAL DETERIORATION.

When those children in turn become parents, they are less fruitful. In the next remove, they are not well developed. Their muscles are slender and flabby, the form inclined to be gaunt, and in mental force they are inferior to those from whom they derived their being.

Besides a physical deterioration, a scrofulous diathesis begins to appear. They hold out longer than hybrids from domesticated animals, with the exception of the mule; but according to the remarks of Mr. Forbes, they actually cease to propagate in the fifth remove from a union of Caucasian and African blood.

## MANLY PERFECTION.

The negro is a man for Africa, the Malay for the East Indies, and whites for temperate zones and hyperborean latitudes. In the temperate, the white man attains the highest condition of which his nature is susceptible.

What is a half-breed? In this country it is understood to be a child of a white father and a squaw. They have never been raised to any prominent positions of usefulness through a native spirit of energy; nor, even when assisted by conscientious, painstaking philanthropists, could one of them be made into a counsellor, a man of thought, of any value to the interests of society.

No educational discipline conducted with special reference to proving their capacity for progress, or how splendidly they may operate as instrumentalities in advancing the civilization of savage tribes, has ever been successful.

They have never gone forward, aided by such facilities as have been urged upon them by Christian charity and governmental patronage, to the achievement of any results, predicted and hoped for by their warm and sincere friends.

Half-breeds may be persuaded to reside in houses superior to wigwams, to cultivate fields, and wear clothing more complicated than a blanket; yet they do not readily fall in with the ways of civilization. They have neither been made scholars nor very devout worshippers. They are just as near to the usages of ordinary civil life as they are to the white man in blood, but no nearer. They have to be sustained by unrelaxing effort, or they quickly deteriorate by running into those wild habits of indolence which are predominant in the nature of the stock from which they came, always stronger on the Indian side than on the other.

Some few individual half-breeds have been rather successful in elementary agriculture. They may raise corn, send their children to schools provided for them, but no scheme has yet been successful in moulding them willingly and heartily into the ways and habits of Anglo-Saxons. They never can be kept up to any standard of civilization to which they have apparently been raised.

### DISAPPEARANCE OF THE INDIANS.

Gradually and inevitably both Indians and their half-breed descendants will wholly disappear from the continent. A few centuries hence there will not be a remnant left of the red race which once roamed with unrestrained freedom, like the game they pursued, on the broad expanse of North America.

Indians are a pioneer race, whose mission is nearly accomplished. Nations were here before them. Millions of human beings, who are only known through the monuments that remain, the evidences of their industry and labors in the rearing of mounds and earthworks, which have outlived the name, the fortunes, and the history of those who raised them, were exterminated by these remnants of powerful invaders, whose gradual extinction is certainly decreed in the court of destiny.

All such revolutions,—the appearance and disappearance of races,—are in conformity to a law of limitation. Nations, like individuals, carry in their organization the seeds of dissolution.

### TENDENCY TO DISEASE TRANSMISSIBLE.

Children of consumptive parents are born with minute tubercles in their lungs, embedded in elastic pulmonary tissues. Their existence may not even be suspected; but when exposed to influences which inflame them, they burst and ulcerate through the delicate air-cells, and death supervenes.

Children of consumptive parents rarely escape the fatal malady. Even if no incipient tubercles are quiescently slumbering in their lungs up to the middle age of life, when reaching the period at which the parents fell under the disease, they are pretty sure to pass away in a similar manner, provided they remain in the same locality.

By taking up a residence where the atmosphere is freer from humidity, vitality may be very materially recruited, and life prolonged. But whenever tubercles are present, as a direct inheritance, no methods have yet been successful in preventing them from inflaming, softening, and degenerating into pus.

When that stage is ushered in, the skill of medical practitioners avails nothing. When those organs in which vitality is manufactured—that is, where oxygen is separated from atmospheric air, and carbonic acid thrown off—are actually destroyed, a recovery is impossible.

## DECEPTIONS OF QUACKS.

Nothing is more preposterous than the vaunted pretensions of those empirics,—criminal quacks who raise expectations, by announcing the restoration of consumptives by new methods of medication, generally their own.

A destruction of the parts of an organ in which vitalizing properties of the air are brought in direct contact with arterial blood, must terminate fatally. No regeneration of destroyed parts can be made by any process within the range of science.

## ACTUAL MALADIES INHERITED.

Scrofula is transmissible; so are syphilitic taints, and some eruptive maladies. The latter are traceable, carefully investigated, quite frequently to the same source. Even a predisposition to deafness, nervous irregularities, distorted fingers, incurvated nails, enlarged joints, St. Vitus's dance, and all shades of insanity, pass from family to family for several generations, rather gaining intensity than losing force.

A tendency to bleed profusely, and even to die of hemorrhage from slight punctures, or the simple extraction of a tooth, runs in some families, without remedy.

## INHERITED PHYSICAL EXCELLENCIES.

Such facts, and many more illustrative of the law of transmission, are familiar to physicians. It is equally true that personal beauty, fine teeth, a tall figure, a musical voice, a mathematical brain, are inherited and propagated, like moral qualities.

## Imperfection of Art in Saving Life.

Surgeons, of extensive experience, have often failed to arrest hemorrhages in one of those so-called natural bleeders. Whether their blood is deficient in that plastic element which assists coagulation, or whether a retraction of the lips of wounds in them, which cannot be kept together by ordinary mechanical appliances, is owing to some peculiar spasmodic contraction of tissues, has not been ascertained.

Compression, styptics, or, indeed, any of the commonly known modes of arresting a flow of blood in those thus predisposed, are ineffectual.

## Selections in Marriage.

It behooves those expecting to enter upon the responsibilities of marriage, to weigh well and investigate a family history before such relationship is formed.

A past and present sanitary character of a family with which marriage is proposed, is of far more importance than might at first be supposed, since various conditions in regard to body and mind are propagated, and may lead to individual sufferings and misery through generations in the future.

Such inquiries, of course, would have to be conducted in a very guarded manner; otherwise, not only much offence might be roused, but the whole matter considered impertinent and ridiculous.

But a regard for one's own comfort in the possible appearance on the stage of life of others for whose well-being, character, and condition, the happiness of parents will be at stake, fully justifies such inquiries and investigations.

If a young lady has ascertained that consumption is a

hereditary malady in the family of the man who proposes himself for a husband, prudence should influence her not to peril herself, or the children she would probably bear, to the contingencies that surround a family predisposed to a lingering and fatal disease.

She could avoid a prospective trouble. It is useless to extend the argument against being joined in wedlock with a man who is certain to die, as his father, mother, brothers or sisters had died, of pulmonary consumption.

### PROGRESS OF PULMONARY CONSUMPTION.

An amazing destruction of human life from that incurable disease is all the while going on in the United States, particularly in the northern parts.

Without regard to the laws of probability, or the destruction of the fair, bright, beautiful, and intelligent, in the beginning of life, by that malady, even to the extinction of families, little or no thought is given to that which is pretty sure to occur when marriage is proposed.

The farmer selects the soundest, best-developed seeds and appropriate soil, otherwise the harvest would be small and imperfect. In the raising of stock, none but the soundest in health, best-formed, and exhibiting indications of constitutional vigor, are allowed to propagate. Thus the high-bred horse, the splendid ox, the finest sheep, and choicest poultry are obtained, by determining from what source they shall spring.

Nature manages among birds and all animals, in a way to secure health and strength, by not permitting the weak, feeble, puny males to generate at all. They are driven away and kept at a distance by the giants of the herd, the flock, and in the poultry yard, who alone are the sires of each succeeding generation.

The female is passive in all those examples, having no partialities or affections to gratify; and thus the blood of each is kept up to the highest requirements of an organic law.

If consumptives did not intermarry, hereditary consumption would disappear. Pecuniary advantages, social condition, and love, each acting with peculiar force, pay no regard to the future, in respect to health.

Children are thus born to linger in pain, and die early. The necrological annals of this nation is a melancholy record. It is not diminishing, but, on the contrary, increasing with the increase of population.

When the celebrated Spurzheim was in this country, he fearlessly declared in public, that the legislature should interpose its authority, by interdicting the marriage of consumptives.

## RISK IN MARRIAGE.

In cities, particularly, ladies hazard more in entering upon matrimonial relations than in the country, where the avenues to vice are fewer, and dissipations, generally, are frowned upon with a severity that inaugurates a better system of morals.

Physicians alone know of the extent of taints which fester in the veins of men in cities, who, perhaps, are envied for their possessions, their social positions, and their influence.

Men are more prone to irregular lives than women. They plunge into dissipations, of which their most intimate friends have no knowledge, and contract diseases, for the relief of which they dare not consult their own physician, as it would expose their doings where their reputation is enshrined in gold.

Quacks tamper with them, get their money, and keep the secret. Being no way qualified for medicating a patient with

grave complaints, the canker that gnaws and undermines their
health is not eradicated, but a poison is left behind, to annoy
and worry the sinner the remainder of his days.

A father, whose system contains the seeds of an eruptive disease, a scrofulous tendency, a syphilitic taint, deep-seated ulcerations, unsound teeth, an offensive breath, from internal causes,
which speak as plainly as such complaints can announce their
existence, will pretty certainly transmit them to his children.

Very many women have contracted diseases from that
source, which have made them invalids, and destroyed all the
comfort of life, without, perhaps, ever suspecting the origin of
their protracted misery.

Cities abound with showy, flashy, fascinating impostors,
and women are their dupes. Fine establishments, fashionable
appointments, and costly equipages, however, are no compensation for the loss of health. When they become the wives of
such men, they are prisoners in a charnel house.

A reformed rake is not the material for making a good husband. It is the privilege of ladies to decide whom they will have;
but unless the candidate for their hand and heart have a character as transparent as glass, it is for their interest to weigh
every circumstance with extreme deliberation, before saying
*yes* or *no*.

### TRANSMISSIBLE TENDENCY TO INSANITY.

Insanity is another transmissible misfortune in families.
Beware of a lover whose father or mother has been a lunatic.
Severe reverses, loss of friends, peculiar affliction, and unfore-
seen accidents, may give rise to distraction. Such forms of insanity are not without hope, when the cause has been removed
that gave rise to them, and should not therefore be viewed in
the same light as a hereditary predisposition to insanity. Nor

need there be an apprehension of a transmission of any temporary cerebral irregularities, the result of such causes.

## Beware of a Predisposition to Intemperance.

A transmitted predisposition to suicide, a murderous propensity, and a morbid craving for strong stimulants, are each of them elements that lead to all imaginable unhappiness. Avoid them, therefore, in a lover.

Ladies sometimes marry men who are known to give manifestations of those fearful conditions, under an impression that they can manage them. To marry an habitual drunkard, when the fact is known, under an expectation of wielding an influence that will lead him to abandon a debasing vice out of respect to a wife's feelings, is an absurdity. They have no powers of self-restraint, nor a wife any influence with a drinking husband.

It is an experiment without a way of escape from impending misery, shame, and degradation, when a lady of refinement weds a dissipated man. It is a cruel wrong when friends match youth, beauty, health, and accomplishments, from sordid motives, to an old, shattered body. It is a fearful plunge into an abyss of misery.

## Wealth Buys what cannot be Won.

Such irrational marriages scarcely differ in moral turpitude from a direct sale. It is a legalized abomination.

Property is the object when a blooming miss in her teens weds an octogenarian. If there were no money to be won by a game of chance—for it is one, in which the bride fully expects the grave will quickly cover up the old carcass she hates—such unions would not take place.

Ambition to be rich urges brilliant women to risk their happiness on a throw of a matrimonial die. How frequently the community is astonished by such voluntary exhibitions of unnatural alliances,—a living woman chained to a corpse.

Where is the tenderness, the sympathy, the religious sense of honor, the instinct of love, when a woman in the vigor and aspirations of youth sacrifices all at the shrine of money?

In commenting on the barbarous customs of the Orient, where females are sold at prices varying according to physical attractions, travellers invariably express their disgust. It is a system which Christian civilization frowns upon with indignation. But are there not sales in the United States, even more extraordinary?

Blue Beards are not all dead yet. Those women in market, waiting for the highest bidder, offering themselves voluntarily, are neither sacrifices, nor ladies. They are beings without heart, without conscience, or a sense of religious accountability to society or their Creator.

A lady is a different being. When her moral qualities and the attributes of her gentle nature act in the sphere where she ought to move, she is recognized as the best gift of God to man.

# CHAPTER XXXVI.

## DIVORCES.

Being Matched—Too Easily Procured—Incompatibility—Progressive Infirmities—Matrimonial Bickerings—Congeniality—Commercial—Children's Society—Companionship, etc.

UNHAPPINESS in marriage is obviously on the increase; lamentably, too, in the highest circles of intelligence in this country.

A direct evidence of this statement is found in the courts of law from the Atlantic to the Pacific, and from the forty-fifth degree of north latitude to the Gulf of Mexico.

Demands for separation from bed and board have become disgracefully common all over the United States. Neither legislators, divines, or moralists, have been successful in keeping the family fold in that condition of contentment, which is theoretically, if not practically, the basis on which rests the institution of marriage.

## BEING MATCHED.

When contracting parties are only paired, but not lovingly matched, they become estranged, most unaccountably to themselves. Divorces do not appertain to any particular condition of life. Clergymen, lawyers, physicians, merchants, bankers, actors, authors, the affluent, the tall, short, fat, lean, and even among the industrious, wealth-producing classes, quite down to cellars under sidewalks, all have their representative dissatisfied applicants for relief from the self-imposed shackles of matrimony.

So urgent is the desire for emancipation, by slipping their necks out of the conjugal noose, enactments are undergoing modifications in several States to facilitate a retrograde progress in Christian civilization.

## TOO EASILY PROCURED.

Divorces are procured with disgraceful ease, to the amazement of those who in other countries have been brought up to hold sacred an obligation to abide by a marital promise, to religiously hold out, for better or for worse, till death doth them part.

A facetious story went the round not long ago, of a Massachusetts man who wrote to the clerk of the Legislature of Indiana, to ascertain why his petition for a divorce had not been acted upon. In answer, the official wrote back it was customary in that body to proceed alphabetically; therefore he must not be impatient, as it would be impossible to reach $M$ till late in the session, as they had only reached $B$ in the regular order of application.

## INCOMPATIBILITY.

A proximate cause of such incompatibility, the generally alleged reason for wishing a dissolution of the bond, is explained upon what is called vital repugnance.

There is a kind of congenital uncongeniality, not to be overcome or subdued by any known process, says a new theorist, because there is a difference in their predestined longevity.

Thus, if a man is twenty years the senior of his wife at marriage, they may possibly sail over life's tempestuous sea with tolerable equanimity a few years.

Going with the tide, however, is not their lot. Both are occasionally rowing against a strong current, without keeping time. Hence the boat is swayed, first, one way, then, in an opposite direction, instead of gaining a peaceful harbor, protected from storms and tempest blasts.

## PROGRESSIVE INFIRMITIES.

After awhile the husband begins to exhibit the infirmities of age. Besides, he has gradually established certain fixed rules which, in his long experience, are considered fundamental principles necessary for repose, for propriety, for happiness; and it very much ruffles and disgusts him, too, if others refuse to conform to the routine of regulations he resolves to establish in his own household.

Madam entertains widely different views of the subject. She comments upon his propositions as either preposterous, ridiculous, or arbitrary. He makes no allowance for more youthful feelings, while the wife, on the other hand, makes no effort to conceal her dissatisfaction in being obliged to humor the caprices of old age.

## MATRIMONIAL BICKERINGS.

With occasional cutting remarks to the discomfiture of both, the spirit of division obtains a foothold. An old husband of a young wife never inspires her with reverence for his bald head or gray hairs. Love never was an element in the original arrangement. Both were deceived in supposing they were made for each other.

An old man may have wisdom, judgment, and a handsome estate, but he cannot inspire love and the warmth of affection in a girl twenty or thirty years younger than himself. She feels no sense of companionship in his society.

While an old husband is deteriorating, and closing gradually into smaller compass, the young wife is developing into the fulness of commanding womanhood.

Reverse the circumstances. The wife being advanced, no longer throwing off those magnetic influences which are the bonds of attraction, a want is felt; but what it is, words cannot properly express. It is a sympathy only to be engendered between those nearly of the same age.

An aged wife, the senior of the husband ten or fifteen years, may be a model woman in the management of her domestic duties, prudent and eminently discreet; yet they do not harmonize, though both are good and true.

When nearly of the same age, their views, feelings, and opinions keep pace on the same vital plane. One rarely acts without the other in anything of importance, or suggests a measure which would not be of mutual benefit.

### CONGENIALITY.

There is a complete oneness with them, when appropriately brought together. That is matrimonial happiness which we read about, but do not as often witness in real life as might be expected in a Christian country.

True unity of soul is the foundation of all the felicity found in marriage. In that delightful realization of what actually belongs to marital relations, of which affection is the bond of union, one party has not a longer expectation of life than the other,—an unconscious harmony which, nevertheless, has a direct influence on their mental and physical organization.

With such a pleasant preparation for travelling together on the highway of coming years, marriage is a divine institution.

Two merchants of nearly the same age agree together far better in their business affairs, than when there is considerable difference in their probable tenure of life—all other things being equal.

## COMMERCIAL RELATIONS.

Some of the oldest and strongest commercial houses were established by youthful partners, whose plans, operations, and methods of conducting their enterprises were results of seeing objects from the beginning alike, because both were alike impressed by the same surrounding influences.

Old capitalists in business rarely proceed so smoothly with a young man as with one of their own age. A reason is sought for in that natural law of correspondence which is recognized in various relations, but which is extremely difficult to elucidate. There is a parallelism in thought, in reasoning processes, and a unity of feeling, in those of nearly the same age. Having lived about the same number of years, they reckon from the same events and epochs.

## CHILDREN REQUIRE THE SOCIETY OF CHILDREN.

Children require the companionship of children. They never establish the same kind of familiarity with grown-up persons as they do with those of their own mental calibre.

Impressions from common objects strike them so differently. The conversation of an infant is insipid to a man of years, while the chat of the latter is totally beyond the comprehension of a little prattler at his elbow.

Domesticated animals, to an observable degree, are influenced by the same law of association. An old ox takes no interest in a calf, but lows at the sight of a distant herd. Old dogs hardly tolerate the pranks of puppies. Cows covet the company of cows, and old singing-birds appear to have pleasure in the society of those similar to themselves.

### COMPANIONSHIP OF ANIMALS.

Some animals form a warm attachment for each other, provided they have been a considerable while together; but they are not particular in expressions of friendship, if they associate late in life.

Coach-horses, after having been accustomed to work in the same carriage, upon being put in adjoining stalls, become excessively uneasy when separated, and exhibit gratification in their whinnyings of recognition in being again harnessed in the old way.

Two cows pastured in the same field, or stalled in the same stable, or two oxen accustomed to the same yoke, exhibit very decided uneasiness on being separated. Their nervous watchfulness, vigilance, and frequent calls at the top of their voice, is a language that denotes the violence done to their attachment to an old friend.

A young weaned colt cares but little for a sedate horse; nor does a spavined hack in a dirt-cart covet the society of antic nags, even when at large in a broad enclosure. Kittens are repulsed by sober cats. They may tolerate their presence; but when they begin to take liberties in their mischievous capers, a growl, or a blow with a sharp claw admonishes them not to presume upon the gravity of their seniors.

### REPTILES WITHOUT ATTACHMENT.

Reptiles do not appear to possess social feelings. Neither do voracious fishes, as sharks, wolf-fish, etc. On the contrary, cod, haddock, mackerel, and many other tribes are sociable, and range in company over their feeding-ground, and migrate in immense armies for mutual protection and society.

Whales are social in their nature, also, as porpoises are; both swimming amicably together in their pastimes, or in pursuit of food.

Whales, after all, are not fish. They belong to the mammalia. They breathe air exclusively, and suckle their young.

When aged men or women advanced in the vale of years marry those younger than themselves by many years, it is not only a gross mistake, but it is also a violation of a natural law. It is as true in social science as in homœopathy, that like cures like. In other words, a condition in age, experience, and force of vitality, is essential to that happiness which is the incentive for assuming the legal and all other responsibilities appertaining to marriage. Discrepancies in those respects are sure to eventuate in certain disappointment and marital wretchedness, where neither one is influenced by a highly developed religious sentiment of accountability.

Who can doubt that the friction of a wounded spirit, chafed and fretted by an uncongenial marriage, must be productive of intensified mental misery?

Who does not believe, also, that where a man and woman of suitable age, of cultivated intellect, refined in character, are lovingly united, they will find all that calm, ennobling realization of their expectations in that relation?

In a felicitous marriage, longevity is promoted, health is better secured, and if heaven is ever found on earth, it is in the home of such a family.

### INDISCREET MARRIAGES.

When a young woman marries an aged man, she perils her health,—possibly, her life. He will improve, because his system will imbibe her vitality. If some ladies have sufficient vitality accumulating, to bear the draft of what may be called absorption of life, a few years, they may outlive the old husband. Ten fall by the way, however, where one survives.

And in those cases, if it could be fairly explained how she escaped the penalty of a violated law, it would unquestionably be due to an estrangement,—protecting herself by not being within the reach or magnetic conducting force of the body which would otherwise have received her vitality.

Reverse the conditions, and a young man would peril himself precisely in the same way.

Such are the mainsprings of life, subtle and incomprehensible, but they are the laws that influence and govern humanity in every country.

# CHAPTER XXXVII.

## The Longevity of Women.

Life a Precious Boon—Modifications of the Penal Code—Experiments—Mind
Independent of Body—Suicide a Crime—Women in their Desperation—
Women Live Longer than Men—Have Better Habits—Life Limitation
—Pulse—Life Insurance Positions—As Cultivators.

A DREAD of death is implanted in every human breast.
Even creeping insects have an instinctive apprehension of fatal
consequences, attending exposures to superior force.

A small animal is in fear of a large one. It is'a feeling
that cannot be overcome, because it is incorporated with their
nature as a safeguard to inspire vigilance for self-preservation.
Otherwise, unapprehensive of impending dangers, and regardless of consequences from a relaxation of that sentinel sense,
they, and man too, with all his calculating faculties, seeing the
end from the beginning, in his reasoning from cause to effect,
would heedlessly plunge into a vortex where certain destruction
was inevitable, as he would lie down upon a soft couch for
repose.

Life is a boon too precious to be neglected, or carelessly
thrown away. It is an imperative duty to live as long as we
can, and in all Christian nations it is considered a crime to voluntarily destroy ourselves or others.

A doctrine is obtaining rapidly, the advocates of which are
already numerous, that God, who gave life, has alone the right
to take it away.

Very marked modifications of the penal code have not only

already been effected through the spreading influence of disbelievers, in the necessity or right to inflict capital punishment, and still further alterations may be anticipated.

Starting with that proposition, relaxing the severities of punishment for several very common crimes within the last few years,—they are not as frequent as they were. A further reduction of legal cruelties, unworthy an age of elevated Christian advancement, will prove a surer remedy than hanging on a gallows.

Extreme cases, characterized by atrocious barbarities, and premeditated wickedness of the perpetrator, should be placed beyond the control of executive pardoning powers. A perpetual imprisonment, wholly and entirely beyond the reach of a governor or a president, would be so terrific as to restrain those who have entertained an expectation of freedom at last, even under a life sentence.

Perpetual incarceration, without the possibility of ever being again restored to freedom, would be dreaded far more intensely by great criminals, than a public execution.

When it was announced to the first murderer that, instead of being put to death, he should live, and seven-fold vengeance be the penalty of any one who injured him,—a mark being fixed on his person that he might be recognized as under an awful sentence, the wretched Cain exclaimed that his punishment was greater than he could bear.

Inquisitive physiological experimenters have interrogated nature with a view to ascertaining whether life departs instantaneously with a stoppage of the vital machinery.

When a person has been shot through the head, heart, or the solar or semi-lunar plexuses in front of the spine below tho diaphragm, does consciousness linger awhile and then gradually take a final departure, or is death instantaneous?

Every muscle has a special life endowment of its own, quite independent of the will. After being lacerated, and, indeed, after being separated from its connections, while there is contractility remaining, there is life in it.

Chemical decomposition is the only certain evidence of death.

The conscious soul exhibits its peculiar properties through the instrumentality of an organized body. In drowning, considerable time evidently elapses before life is extinct. Remarkable cases of suspended animation incontestably prove that, if the soul had departed on its never-ending mission to eternity, it was actually recalled back again by the appliances of art.

In drowning, the union of body and mind is gradually dissolved; but it may be interrupted and death prevented, by manipulations that set the vital machinery again in motion.

The mind, therefore, is there for a while; and it is probable the same condition exists in decapitations. But violence inflicted on those highly-vitalized instruments by which it manifests itself in life, is a shock dissolving instantly the connection between body and mind.

With the escape of arterial blood in a gash, in cutting suddenly through the neck, the brain is deprived of the material it must have to act at all; and hence death speedily follows, though there may be an instant or two of distinct consciousness. The Paris *savans* represent a decapitated head as engaged in thinking for a short space of time, deprived of the ability of expressing its wishes.

There are conditions in which all are cowards. Men may fight bravely, face the king of terrors at the muzzle of a cannon; but, when raising a weapon for destroying their own lives, it is with fear and trembling.

If ever suicide is accomplished with a firm, unflinching will and a steady hand, it is charitable to suppose the individual absolutely insane; because the act is a notorious violation of the strongest instinct of his nature.

Women in their distraction wildly perform deeds of desperation against themselves. They leap into abysses of misery to avoid a foreseen disgrace. Nothing so nerves them to face the dreadful alternative of death or shame, as questioning their moral purity. What is life to them without the consciousness of unsullied virtue?

That is the question with a woman nurtured in a religious belief of rewards and punishments in a future state; and hence, among professing Christians, examples of self-destruction of females are more common than in pagan or Mahometan countries.

Pagan and Mahometan women rarely commit suicide. Education, therefore, shapes the mind, and plants deep down in the recesses of the heart, those principles which both govern and direct them in their social intercourse.

Whatever is instilled into the mental constitution of the girl remains there through all the meanderings of after years: so certain is it that, as the twig is bent, so the tree inclines.

Inquiries into the physical signs of longevity in man fully confirm the opinion that women, on an average, live longer than men.*

They are less exposed to dangers which sweep off men and

---

* Professor Faraday has given it as his opinion that all who die before they are a hundred years old, may be justly charged with self-murder; that Providence, having originally intended man to live a century, would allow him to do so if he did not kill himself by eating unwholesome food, allowing himself to be annoyed by trifles, giving license to passion and exposing himself to accident. The French *savan*, Flourin, advanced the theory that the duration of life is measured by the time of growth. When the bones' epiphysis are united, the body grows no more, and it is at twenty years that this union is effected in man. The natural termination of life is five removes from the several points. Man, being twenty years in growing, lives,

boys at sea, in armies, mines, manufacturing establishments attended with perils, as the making of gunpowder, explosive cotton, nitro-glycerine; and in various circumstances of peculiar contingencies, to which females are rarely, if ever, subjected.

Then again, women, as a body, always have better habits, and better morals,—a sentiment often repeated in the pages of this volume. Their vitality is not wasted in midnight carousals, nor are they guilty of enervating vices, which kill off men frightfully fast, of which little is known out of the confidential circle of the medical profession.

They commence life under more favorable circumstances, in some respects, in regard to the preservation of health, which thousands of them fritter away prematurely, in coming into womanhood. Still, more women live to very old age than men.

An examination of a family's necrological record, if carefully kept, discloses some curious facts illustrative of the tenacity of life in females who have escaped the tortures imposed upon the fashionable sisterhood.

Constant practice in the examination of applicants for life insurance has enabled medical examiners to arrive at certain interesting conclusions respecting the death period of women, which had escaped notice before those investigations were instituted.

Limited as may be our knowledge of vital force, enough has been ascertained for the construction of tables of expectances. That is, if a person has arrived at any particular age, it is expected he or she may live a certain number of years from that date.

Physicians make mistakes in their estimates of the value of life, as well as others not supposed to be as well informed in

---

or should, five times twenty years; the camel is eight years in growing, and lives five times eight years; the horse is five years in growing, and lives twenty-five years, and so on with other animals.

regard to the probabilities of life; but they are progressing, becoming more critical, and more accurate also, in their investigations. Their ears and the sense of touch are being educated with reference to discriminating, with precision, between normal and abnormal sounds of the heart. When careful in their examinations, it is surprising with what success and readiness they detect irregularities in the circulation, that would escape the attention of those inexperienced in those pursuits.

Unfavorable conditions of the heart, the lungs, kidneys, etc., require a very nice perception of variations from their action in health or disease.

Those organs are exceedingly over-worked, and, therefore, driven into a degree of unnatural activity by the habits, bad customs, imagined business demands, and vices of the times; and the consequence is, an increased mortality from those sources, quite rare among our old-fashioned ancestors, who proceeded with moderation in their affairs.

A hurried pulse, far above the ordinary beats of the heart, when the expenditure of vital force is in equipoise with the ratio of supply, tends to injury. We are constituted for excitements. If not too long continued, no injury accrues. But when the tension is kept up continuously, too long, the next phase is debility.

A preparation for being examined for a life policy sometimes quickens the pulse exceedingly; and one not accustomed to the sudden changes which emotions of the mind may produce, is liable to grave mistakes.

Medical gentlemen are occasionally blamed for mistakes, as though they were, or at least ought to be, infallible, when in the service of insurance institutions, trust companies, and the like, where professional opinions are required in granting their favors.

Those who have passed through a tedious professional preparation for being eminently qualified to discriminate one sound from another, or determine by pressure at the tip of a finger, whether an applicant for life insurance may live ten years, or expire, in all human probability, in ten days, are attended with painful anxieties, and environed by more responsibilities than officers of such institutions recognize.

Unfortunately, for the honor of the medical profession, the qualifications of medical officers are decided upon by persons who have not the requisite knowledge of the value of science for guiding them in a choice. A blockhead is quite as often chosen to a responsible professional position, as a man of superior attainments. A pecuniary influence, or relationship to one of the directors or an influential stockholder, may decide an appointment. Is merit ignored? This declaration is abundantly sustained in looking at the names of some who are the best bowers of many life-offices, but who could not pass an examination for the position of a village pedagogue.

Admitting the capacity of women for occupying all places, and for engaging in almost all pursuits heretofore considered the special properties of men, the further we proceed the more openings seem to present for them.

It is not desirable that they should unsex themselves for the sake of employments which in ages past have been denied them. It is by no means necessary that they should ride a horse like a moss-trooper, tend saw-mills, hew stone, labor in quarries, coal-mines, iron-foundries, or anywhere in which their presence would be inappropriate.

Women may be admirable gardeners, florists, fruit-growers, wool-raisers, cultivate vineyards, or, indeed, as many of them do, carry on extensive farming operations. Fruits have always commanded good prices; the demand, thus far, has been greater than the supply. They are wanted everywhere.

Having a natural aptitude for horticultural industry, and a delicate taste in selecting and directing, what fortunes are in reserve for those who early embark in those employments !

Berries, poultry, honey-bees, piscatory economy, all of which may be conducted on a few acres of ground, present an inviting field for the display of female energy, enterprise, and praise-worthy example.

Women need not necessarily lose caste among the refined of their own sex, become rough in manner, or demoralized by coming in contact with mother earth. Their figures will neither become gross, their features less attractive, their charms deteriorate, or their beauty fade any sooner for identifying themselves with the culture of fruits, flowers, wheat, or wool.

. People will have luxuries if they go without necessaries. Among profitable pursuits for females, requiring neither un-pleasant associations nor hard labor, is honey-making. One woman could easily manage one hundred hives. Even one, not occupying a square yard of ground, would supply thirty, forty, and up to sixty pounds of honey in a single season, if carefully superintended.

Let one thousand females embark in apiarian enterprises any-where, and without the least regard to the quality of the soil or its capacity for yielding flowers. Bees collect honey from great distances, and store it wherever we direct. What wealth would be accumulated by that one thousand operators in honey !

There are more flowers in yards, windows, open conserva-tories, parks, and highways in most cities, than in four times the same area of land in the country. Even if there were not, a single flower within five miles of an apiary would be found and regularly visited by city bees. Therefore, rear them in cities if the cultivator has a stationary home.

The experiment has been tried, and crowned with entire

success, even when the hive was kept within a cool part of the building, and the foraging insects went out and in through walls.

Throughout the country there is not a poor widow, a forlorn spinster, or an idle woman, who cannot descend to pursuits below the estimate she places upon her social position, who might not have a pleasant revenue from this delightful employment of honey-raising.

Throughout continental Europe, peasant women are accustomed to labor in the field side by side with men. They have the same organic structure, functional peculiarities, instincts, and necessities of the most elevated of the sex, yet they are quite overlooked in researches for physical signs of incapacity for such lives as they lead.

Because women can endure hardships, can labor, lift, dig, saw, carry burdens, and drive teams, it is not an argument in favor of obliging them to do so, neither does it accord with our civilization not to attempt relieving them.

Studying their condition at every step, from the lowest to the highest round in the ascending ladder of life, in all countries, the conclusion arrived at, in reference to their longevity, is this: that more women live beyond a century than men, their circumstances, *ceteris paribus*, being equal. Both, however, in communities most distinguished for culture and intelligence, fall far short of the years they would have attained to, had they not violated many immutable laws of health.

Public registers abound with notices of men and women who have lived far beyond the supposed ordinary limit of human life, on the presumption that threescore and ten is the doomed measure of our days.

The more quiet, unobtrusive, and less exposed way of life of women is favorable to their longevity. They are rarely sub-

jected to those sudden assaults upon the constitution, those fric-
tions of a rude world, or those personal contentions, which
wear away men. As they are rarely exposed to storms, or
called upon to test the strength of their muscles, or perplex
their brains with problems and difficulties which break down
strong men prematurely, their chance for prolonged life is
better.

Individuals pass through dreadful trials of body and mind,
and thousands of females throng madhouses, the victims of
cruelty and oppression; but, as a whole, the expectation of
life is altogether in their favor.

Women are less corrupt than men, even when wickedly
debased by vicious associations. They think less evil, avoid
polluting influences, and thus are secured from many direct
causes of premature death.

Were it not for wandering too far into the regions of
antiquity, illustrations of prevailing opinions, that females had
a peculiar tenacity of life, might be gathered. But we ascribe
what in the olden time was thought an extraordinary endowment of vitality, to their habitual sobriety, propriety, and
happy exemption from turmoils and excitements that wear out
men.

The book of Genesis gives a narrative of the old age of
Sarah, and a remarkable physiological revolution in her system,
perhaps hardly ever paralleled since. Becoming a mother in
extreme old age, is by no means a common occurrence.*

---

* Mr. W. J. Thoms' new book, " Human Longevity : Its Facts and its
Fictions," demolishes the pretensions of many of the marvellous " old men "
of tradition to have lived a century and upward. He clearly proves that
" Old Parr," Jenkins, and the Countess of Desmond, who are reputed to have
survived to 140 or upward, are cases of longevity resting upon no positive
evidence. He demonstrates that the ages of a more modern series of centenarians were as follows :—Mary Billinge, not 112, but 91; Jonathan

Erythea, the Sybil, says that Phlegon lived ten hundred years. In the writings of Matthew Paris, it is asserted that the Wandering Jew was recognized in 1229. Next, copying the story of a man who, at the age of three hundred and thirty-five, was brought into the august presence of Lopez de Castenada, while viceroy of India, and similar extravagant legends, a formidable array of strange biographies might be collected. They are of no value, being curiosities of history, once believed to be true, when a few monks wrote for the astonishment of ignorant, superstitious millions.

Whether persons in modern times have attained patriarchal longevity, admits of a reasonable doubt. However, there are exceptions to general laws; and whenever a man or a woman passes beyond one hundred years, their vitality must have been remarkable.

Henry Jenkins, who in his twelfth year led a horse laden with arrows to the battlefield of Flodden, reached the age of one hundred and sixty years. This case seems well authenticated, as the record of his burial by a national subscription gave extensive notoriety to his extraordinary longevity.

Thomas Parr died at one hundred and fifty-two. He was buried in Westminster Abbey. That circumstance shows the deep public interest in the fact that he had reached an age that created universal surprise, and, therefore, he was entombed within the sacred edifice, where none but memorable persons were honored in death.

Reeves, not 104, but 80; Mary Downton, not 106, but 100; Joshua Miller, not 111, but 90; George Fletcher, not 108, but 92; George Smith, not 105, but 95; Edward Couch, not 110, but 95; William Webb, not 105, but 95; John Dawe, not 116, but 87; George Brewer, not 106, but 98; Mary Hicks, not 104, but 97. Besides these, a few other cases are introduced, of which all that the author can show is, that there is no convincing evidence of the asserted age.—He admits only two as proven out of the long roll of newspaper centenarians. We have facts to confute Mr. Thom.

When Philip d'Herbelot was one hundred and fourteen, he presented a bouquet to Louis XIV. on his birthday. · "What have you done," asked his majesty, "to have reached so great an age?" The old man replied, being then a government pensioner, "From the age of fifty, please your majesty, I have shut my heart and opened my cellar."

While the National Assembly of France was in session, October 23d, 1789, a man at the age of one hundred and twenty was announced. All the members rose as he entered. Amidst a whirlwind of applause he walked to an arm-chair in front of the secretaries. He then presented a certificate of his baptism, proving his birth at St. Sobin, October 10, 1669. By manual labor he had supported himself till the expiration of a century. A pension of two hundred livres per annum was then granted by the king. A contribution was voted him, and the old man was lodged at the public expense in the Patriotic School. Pupils of all ranks waited upon him.

When Napoleon I. was first consul, he decorated two men on the same occasion, who were one hundred years old, before an immense concourse of people.

At an inauguration of an equestrian statue of Louis XIV., at the Palace of Victories, Aug. 20, 1822, Pierre Huet, called the father of the French army, was present,—being one hundred and sixteen. His countenance was venerable, his voice sonorous, and a flowing white beard gave dignity to his appearance. He had been a cotemporary of the king, whose reign the Bourbon dynasty were commemorating.

Dr. Barnes, of Edinburgh, gives a narrative of Robert Boman, in the *Philosophical Journal* of that city, who lived to be one hundred and fifteen. He distinctly remembered the rebellion of 1714. Among other curious recollections, he remembered when barley was three shillings for a Carlisle bushel;

oats, eighteen pence; butter, three pence a pound; and eggs, a penny a dozen. His food had been principally milk, but he partook of whatever food was prepared for the family. Neither tea, coffee, or tobacco was ever used by him. When hungry, he ate; retired early to bed, when sleepy, but had no fixed habits.

The foregoing cases of extreme longevity have been cited to show that well authenticated cases are numerous, of life being prolonged beyond a century, which has been questioned very frequently of late. But a few only, one in millions upon millions, have had such vitality.

We could show, with equal certainty, that more females have reached an uncommon longevity than males. A very few have considerably passed one hundred years, one hundred and ten, one hundred and twenty; and the Countess of Desmond, one hundred and forty-two!

In the course of one century, one man in many millions may arrive at the age of one hundred years, while within the same period more women live to a full century than men.

Among the Pension Office records, at Washington, on a list of twenty-one surviving soldiers of the Revolution, a few years since, eighteen of them had reached 100 years, and upwards. There were five who were passed 101; four, 102; two, 103; two, 105; two, 106, and one, 109.

In the catalogue of widows of revolutionary soldiers, drawing pensions, there were twenty who were 100 years old, and eleven who were 97. Dinah Vick, at Nashville, Tenn., died 1871, at the age of 109. The oldest person drawing a pension was Chloa Flatford, Virginia, who died at 116. The next person was Charity Flindman, of West Virginia, who was 112.

The Southern States offer more examples of extreme longevity than the Northern. Mrs. McDonald, of Tennessee, died at 106; Mrs. Shaw, of New Orleans, at 107; Mrs. Thrasher, of Georgia, 103; Mrs. Trucker, of N. Carolina, 109, and Mrs. Harris, of Georgia, 101.*

Were it possible, at this moment, to gather a catalogue of the oldest persons, now living, the largest number of the whole would probably be females.

No particular system of diet is superior to another, according to the histories of those who have had such long leases of life.

There is another chronicle of men and women who have violated all the common laws of health, having been as irregular and erratic as the wind, whose longevity equalled those who conformed to all the requirements of a well-regulated life.

Donald McDonald, about forty years ago, was sent to the House of Correction, in Boston, because he was intemperate and quarrelsome, being then one hundred and seven years old! His father died in Scotland at the age of one hundred and thirty-seven, in consequence of an injury to the spine, by falling down stairs.

Who can doubt the transmission of vitality or vital force from parent to child?

Undoubtedly, the death rate is accelerated by intemperance in this and all other countries. We are already called a nation of drunkards, by those who have not had the good fortune to become acquainted with the best specimens of American society. The vice of intemperance is deeply rooted in the constitution of so many, that its baneful and destroying influence taints the blood of those who derive their being from such polluted sources.

---

* Huger, a colored woman, recently died near Alexandria, Ky., at the age of 122. She was born in Virginia, March 21, 1751. She had been blind twenty years. Her memory was good. She was presumed to be the oldest person in the United States.

Neither legislation, moral suasion, nor temperance reformers · have gained much success in their efforts to stay the progress of that dreadful vice. There are temporary lulls and loud expressions of enthusiasm when some newly-devised schemes for reform are proposed. Alas! neither the beauties of sobriety, nor the horrors of a prison make any lasting impression on the case-hardened brains of inebriates. Their morbid thirst is a disease beyond the resources of medicine, in its present imperfect state.

There are persons so organized, they crave unusual excitement. Their temperaments cannot be kept in equilibrium with cold water. They will hazard reputation, and even life, for indulgence. The sober-minded are taxed for the support of vagabonds and criminals, who were made such by intemperance.

There is one untried remedy. When mild wines are cheaper than beer, ale, whiskey, and brain-crazing cordials, those who look upon wine as a luxury beyond their reach will prefer it. Then intemperance will be less frequent, and a new condition of society may be anticipated, most gratifying to philanthropists, to Christian laborers, in the midst of wide-spread vice and dissipation, but not before.

Wine can be made in sufficient quantities to root out the undermining destruction of distilleries. California, alone, has more than sufficient for the remedy—but it must be cheaper.

Give those who are maddened by strong potations something superior to meet the demands of a morbid craving, and in fifty years the public sentiment will sustain this proposition.

Looking to an extended tenure of human life, by conforming more understandingly to the laws of health, which are being better understood through the active influence of the press, and when women rise to that social elevation they are to have, longevity will not be regarded as a wonder, as it has been, but a natural consequence of conforming to those principles which science demonstrates to be the way to health and long life.*

---

* The last census presents the following curious facts :

The total population of the country is about 38,250,000.

Total number of deaths in the current census year, 492,263, or about 1,349 *per diem.*

March seems to be the most fatal month, leading all others by about 1,000. March, April, and May form the most fatal quarter, exceeding any other three consecutive months by over 13,000.

The births number 1,100,475, or about 3,000 *per diem.*

The blind number about 20,000.

The deaf and dumb about 16,000.

The idiotic about 24,000.

The insane about 37,000, nearly one-third of whom are of foreign birth.

Persons over 80 years of age number about 150,000.

Persons over 90 years of age number about 7,000.

Persons over 100 years of age number about 3,500.

Of those over 90 years, the females are in excess by about 1,200.

Of those over 100 years, the females exceed the males by about 1,000.

# CHAPTER XXXVIII.

## Their Future in the United States.

Women Considered in Law—Political Status—Mixed Schools—Vulgar Men—The Evidence of Low Breeding—Schools for Separating Girls and Boys—Russian Apprehension—Annual Teapot Tempests—School Improvements—Political Equality of the Sexes.

In every period of human history, women have been considered inferior to men. All laws for the regulation of society have invariably been so framed as to perpetuate the absurd idea, that they have neither capacity nor a right to participate in concerns of common interest, which tradition, custom, and the sovereign power exclusively confide to male members of the community.

Among savages and barbarians, where that theory must have originated, females are estimated as necessary appendages, but not equals. Before the spread of Christianity their condition, even under the most favorable aspect, was that of slavery. All that has been accomplished for their elevation, and protection in the enjoyment of privileges,—such as they are, even in the most enlightened states of Europe and the United States,—is due to the influence of Christianity.

Women will rise higher, and be sustained in their claims to an equal share of whatever contributes most to the security, happiness, and progress in the world, in proportion to the heed given to those divine truths which were revealed and inculcated by our Lord and Saviour.

There have been fortunate individuals among women in all ages. Some were born to renown. Very few of them have reached political distinction by the mere exercise of superior skill or intelligence,—and simply because barriers were interposed between them and the objects of their ambition by the craftiness of men. When they have held the reins of government by hereditary claims that could not be set aside, they have invariably exhibited qualities that incontestably proved they were not inferior to kings.

Notwithstanding the Queen can do no wrong under the Constitution of Great Britain,—the laws of England take cognizance of a fearful array of illegal acts which her female subjects may commit, and are punished for, without troubling men of low degree, who are guilty of betraying them and destroying their prospects and happiness.

A husband may cruelly abuse his own wife, bone of his bone and flesh of his flesh, and yet she has no redress, or next to none, in one of the most enlightened countries on the globe. How much better is it here?

What does it amount to towards securing better treatment, by putting an infamous, drunken husband under bonds in the sum of seventy-five cents, to keep the peace, who returns home from the police office and breaks his wife's ribs in revenge for having his conduct exposed in court?

A woman is by law held to be inferior to a man in this blessed American Union. While single, she is manager of her own property; but the moment of entering upon the holy state of matrimony, her individuality is lost. She is then Mrs. Nobody in all legal transactions.

As a spinster, she is compelled to pay taxes, can be assessed for public expenditures which do not meet her approbation, but she is not allowed to vote for officers placed in authority over her, nor is she eligible to any place or position of honor or trust under the law.

As a married woman, she loses the privilege she had before of deeding away land, making a will, of doing anything, in short, which her lord and master forbids. On becoming a widow, she regains her suspended privileges, and then buys, sells, and bequeathes her estate to whom she chooses.

Women, certainly, have legal rights here, as wives, but they are very much mixed and obscured by lawmakers, who thus far have contrived to keep the balance of power in their own hands.

In Turkey, a wife is property. If she does wrong, the husband, as proprietor, is notified, he being accountable for her acts. There are no tedious investigations into the particulars of the wrong-doing of which she is accused. If, in his estimation, punishment should be inflicted, he is both jury, judge, and executive officer. The public sympathy is never excited when a member of the harem disappears, if the fact is noised abroad, because it is nothing that concerns the public. A Turkish gentleman may sink a woman once a week in the Bosphorus, without disturbing the placidity of her acquaintances or his own.

Women are worse treated where the standard of moral accountability claims to be far above that of the Koran. Turks are careful of their property. They neither beat, bruise, nor maim their wives. They put them to death when the law requires them to protect the community against a repetition of a crime of which his property,—his woman, is guilty. That is government economy, and saves the expense of a public execution.

### POLITICAL STATUS.

We believe a woman is man's equal, and entitled to all privileges and immunities accorded to men by laws made for monopolizing what by nature and the eternal, unchangeable principles of justice, belongs to her, in making fast their own.

That doctrine is gaining ground rapidly. Until their rights are restored, of which they have for fifty centuries been deprived, civilization cannot be thorough and complete. The millennium will commence when that great day of doing as we would be done by is ushered in, by acknowledging that women are entitled to life, liberty, and the pursuit of happiness accorded to men, whether they have brain, capacity, moral integrity, or not.

Under no ancient or modern form of government have they ever had accorded to them the privileges to which they are entitled. They are of as much importance as men. They were created for one another, and must associate. Being equals by nature, women should share equally and equitably with men in all things. It is only in a republic that a prospect of restoring to them that which has been taken, can be expected. It is rank hypocrisy to boast of republican equality when one half the population is exercising all the power, not even permitting the voice of the injured party to be heard, when the plea is simply in accordance with the Constitution, that taxation and representation shall go together.

We are examining a fundamental principle, not because it would be a gratification to see half the members of the House and Senate at Washington old ladies, alternately taking part in debates on great national questions, and then taking snuff. There are not many women qualified for those arm-chairs in Congress; and if there were, but few would consent to be associated with such rude, boisterous, uncouth specimens of unpolished humanity as are seen there, singularly contrasting with gentlemen of talents, learning, and polished manners, who can rarely be elected.

Women can have equal rights with men, without supposing that every one of them will corrupt voters to get themselves elected a judge, a member of the City Council, or chief of police. Their characteristic honesty and good sense forbids the idea, that they would stoop so low as to accept places notorious for the corruptions of those usually occupying them in these degenerate days of political integrity.

## MIXED SCHOOLS.

Common schools, those elementary, free institutions which are the pride of the people, the nurseries of the national mind, fountains from whence flows a current which both develops and fertilizes every order of intellect, must be sustained, if our liberties are to be preserved.

Throughout the interior of the country, common-school pupils are of both sexes. Boys and girls meet on a common level, pursue the same studies, and stand in classes together. Such are called mixed schools, and they are always the best, distinguished for the progress of the scholars, and the care with which they are governed.

Boys by themselves, or girls in schools exclusively for them, are neither so successful in study or discipline, as when they are associated in their educational pursuits. Mixed schools exert a happy influence on the sexes thus brought into relations which refine their manners, improve their deportment, and lay the foundation for that courtesy, politeness, and civility, which gentlemen, with the slightest pretensions to good breeding, always manifest in the presence of ladies.

When a man walks the whole length of a church without removing his hat till he enters the pew, it is quite certain such

vulgarity is in consequence of never having had the civilizing
eyes of girls in a school-room on him, or a sister or mother to
explain to him that such contempt for the usages of society re-
dound to his injury. If another strolls through a parlor, in the
presence of ladies, with his hat on, assuming an air of independent *nonchalance*, it is another ordinary phase of vulgarity
common with men who had no early advantages of female
society.

Those coarse, foul-mouthed specimens of ignorance and
presumption, observable in men who pretend to consider
women their inferiors, simply because they are not in pantaloons, are to be commiserated for having had no advantages
accruing from female society in their early years, when impressions would have been lasting.

They are the men who have disagreeable wives, deserving
no others. Without circumlocution, it may be assumed as true,
that, by associating while young, boys and girls improve and
polish one another. In large families of brothers and sisters,
they are usually far in advance of those only sons, or only
daughters, who are more remarkable for extreme selfishness
than kindness, suavity, and consideration for others not of their
kith or kin.

Unfortunately, the plan of separate schools is prevailing.
It is a mistake. The old system is the best, and the children
educated in mixed schools will have the best culture and the
best morals.

This view is beginning to be entertained in the higher
seminaries and universities. Young ladies now enter colleges.
The idea of educating a woman as men are educated, would
have been ridiculed once as preposterous. One serious objection to permitting them to enter, was the shocking demoralization and scandal that would follow, from allowing pretty girls

and sophomores to meet in the same recitation-room together, attend the usual exercises under professorial instruction, and then graduate bachelors,—since no charter provides for a degree not conveying to the possessor distinctions belonging to the masculine gender.

All those antiquated objections have been overturned by the good sense and intelligence of a new generation, immeasurably in advance of the buckram of fifty years ago. But there they are, model students, above reproach, and bright examples of what a woman may attain to in the loftiest regions of literature and technical science. So far from exerting, by their presence, a bad influence on frivolous undergraduates, decorum is insured where formerly there were boisterous displays, and industry, where there was formerly inattention and idleness. Young college ladies are a blessing, because order, civility, and politeness are in the ascendant when they appear.

We feel much pleased to express, here, our public recognition of their utility at college in arresting the waves of profanity, cant expressions, and innuendoes that become epidemic where young men are exclusively by themselves, however well reared at home. Those indolent youngsters who used to graduate blockheads, will diminish in number by the admission of female classmates. They would be stimulated to far greater effort, rather than be eclipsed by their accomplished, fascinating inferiors, as women were formerly considered. With all the outcry against the claims of women to political equality, and the spirited determination of the strong-minded representatives of the feminine order for a clear way to the polls, there is not the slightest danger from granting them all they ask. Not one in a thousand would exhibit the slightest ambition for positions they were not abundantly qualified for sustaining creditably to themselves and the honor of their constituents.*

Russia is a despotism. Any aspirations there for privileges corresponding with the developing intelligence of those female students, seem to have alarmed the watch-dogs of the government. It is not at all probable they conducted themselves improperly, or abused their educational privileges in the slightest degree. If the Minister of Public Instruction is excited by apprehensions from the chatty hilarity of a few pretty misses at recess, one such anniversary meeting of antiquated spinsters in green, spectacles, wigs, and bloomers, as proclaim their solemn resolution to lose the horse or win the saddle in New York and Boston, would shake that frozen empire from its centre to its circumference. But here, where we are used to annual explosions of threatened destruction from wind-bags, they only create merriment. The people laugh, the Administration laughs, the reformers laugh, also, and then the tempest in the teapot, concocted by a few dozens of old maids to allay their nervousness, subsides, to reappear the following season, as hybernating animals awake from a winter slumber.

---

* The Russian Government has just made a remarkable announcement in its official organs relative to the Russian women-students in the University of Zurich. During the last two years, says this document, the number of young Russian women who study at Zurich has rapidly increased; there are now more than a hundred in the University and Polytechnic School in that town. It appears that recent developments indicate that these women-students are politicians, revolutionists, radicals, and inclined to free-love, becoming by reason of these things dangerous alike to society, morals, and the government. The royal announcement, after reciting many of these facts, concludes thus : " In order to put an end to this abnormal state of things, it is hereby announced to all the Russian women who attend the lectures at the University and the Polytechnic School of Zurich, that such of them as shall continue to attend the above lectures after the first of January, 1874, will not be admitted on their return to Russia to any examination, educational establishment, or appointment of any kind under the control of the government."

Young women have gained admission to medical colleges and hospitals, both here and in all the principal institutions of that kind in Europe.' There was a prodigious outcry against such impropriety at first. Doors were closed against them by the faculty. But the public sentiment was stronger than the obstinacy of fossilized professors, and the law compelled objectors to give way to the progress of useful knowledge, to unbar the gates and let them enter, sit and learn.

Next, as old cocks crow the young ones learn, says an ancient proverb. Half-fledged medical students pretended they felt themselves insulted by the presence of young female students, whose purity of character, ladylike deportment, and superior culture were a reproach to their own unconcealed coarseness, rudeness, and vulgarity. Time soon corrected their impressions of the deteriorating effects of the quiet attention of feminine intruders, as they saw, to their extreme mortification, that the despised new-comers entirely outstripped them over the course,—won distinguished honors, and left those self-righteous, complacent donkies in the rear.

Precisely the same conflict, to some extent, has occurred here. Both colleges and medical schools have fought bravely in a bad cause, to prevent women from participating in educational advantages which have been too long exclusively considered the birthright of men. The result has been to bring into existence medical colleges for women, and more are required. With constantly increasing numbers of students, the demand for their professional services the moment they are qualified, is opening the eyes of the exclusives. Quite too late for retrieving lost opportunities, with a certain prospect of being outnumbered in attendants, before many seasons have passed, they are now relaxing,—unbolting here and there a door. Colleges are all discussing the policy of submitting to what is inevitable,—the admission of female pupils. Medical schools are losing what they might have secured,—the honor of educating women to assume higher and nobler positions to which they are called by the voice of the people.

## School Improvements Suggested.

As no two persons precisely resemble each other in expression, so they differ in their mental capacities. That school for girls will be best which recognizes this fact by providing liberally, as circumstances will allow, for developing and directing the predominant faculty of the pupil.

Reading, writing, grammar, geography, elementary arithmetic, with some few studies besides, comprises a common school education. In cities, where resources are greater than in the country, singing is taught; sewing, and, indeed, other branches may be taught, supposed to be most necessary for qualifying pupils for duties that may devolve upon them in adult years.

Some children, with slight instruction, would excel in drawing, others in modelling, their organs of imitation being exceedingly active,—craving indulgence. Instrumental music, too, should be systematically taught. There are hundreds of girls in public schools whose genius remains buried forever, just because no proper stimulus to development was ever presented. Musical instruments freely distributed among those who have a taste for music, accompanied by daily instruction from a competent teacher, would bring to light many to become distinguished performers. It would qualify poor girls to rise socially, to earn more with less hard labor, than would otherwise be their lot. Every faculty God has blessed them with, should be cultivated. That is what common schools ought to do. When the rich are distributing wealth they cannot carry away to a world to which they are hastening, rather than give to institutions burdened with funds, why not direct a thousand orders for pianos, harps, accordeons, music books, violins, guitars, etc., to district or other common schools, with an express condition they are for the use of poor female scholars, to qualify them to become instructors? That would be a true specimen of Christian benevolence.

Unfortunately for the world, brilliant talents which the possessors were unconscious of possessing, often remain undeveloped through life, simply because no opportunity for their exercise was within reach of the individual.

No calculations can be made of the amount of buried genius that might have been roused into activity with proper appliances in early school-days, with systematic assistance for bringing it out. It is a duty to assist, to the extent of our means, in the cultivation of all the powers with which girls and boys are blessed. Without aid, scores struggle on, displaying extraordinary natural gifts that cannot be utilized, because imperfectly educated. They know too much of what is unavailable, and not enough in perfection to be instructors in branches for which they have a strong natural bias.

Singing is occasionally taught in a few schools which are so fortunate as to be under the care of gentlemen and ladies who appreciate the importance of having all such branches taught as may be turned to a practical purpose. Some admirable vocal performers have had their musical talents discovered in those exercises, who are now receiving large salaries in church choirs.

Let it be remembered, that a large majority of all the children in all the States never have access to other educational institutions. Therefore let them have all the attention

in common schools that their claims entitle them to as public
beneficiaries.  While taking lessons in reading, writing, grammar, arithmetic, etc., as an agreeable recreation, they could be
taught to play musical instruments, and a conversational familiarity with some language besides their own.  Severe study,
long, tedious recitations, committing to memory what they
cannot comprehend, is not contemplated in this scheme for improvement in mixed schools.  On the contrary, let them as
children learn language by the ear, not by grammatical drillings.
Music must not be taught in that way, because they could not
read notes if they were not carefully taught the value of each
character representing a sound.

## POLITICAL EQUALITY OF THE SEXES.

From a close examination of the great question of the day,
whether women ought to enjoy the political rights and privileges which men exercise, we have arrived at a conclusion they
are quite as capable as men.

Four millions of colored people were emancipated from
slavery, and all the males above twenty-one years of age became
voters *instanter*.  Now, would there be more risk in granting
the same political privilege to intelligent, cultivated women?

Emigrants from foreign countries, utterly ignorant of our language, and certainly so of the constitution and laws of the United
States, in a few months after their arrival become freemen,
voting, and may be voted for; and yet one-half of the native
population, whose patriotism, interest, property, and prayers
for the land of their birth cannot be questioned, are resolutely
kept under control by the law, as not being as worthy to be
intrusted with the franchise as ignorant foreigners, half-breed

Indians, and negroes who can neither read, write, or know the letters of the alphabet.

Wherever the sexes mingle, in the family, primary, common schools, at college, medical institutions, and in society, there is most refinement, courtesy, and intelligence. One more ascending step would place women where they would have political equality, or civilization and the genius of Christianity cannot progress.

If a concession is to be made to them anywhere, if men are ever honest enough to acknowledge the claims of women to equal rights in the pursuit of life, liberty, and happiness, the crowning event and the glorious triumph will take place in a republic : and God grant that the honor may belong to the United States of America !

The late John Stuart Mill, who dared to speak in favor of the elevation of women to higher responsibilities than the jealousy of men in the old world are disposed to permit, thus reasons :—

"That every step in improvement has been so invariably accompanied by a step made in raising the social position of women, that historians and philosophers have been led to adopt their elevation or debasement as, on the whole, the surest test and most correct measure of the civilization of a people or an age. Through all the progressive period of human history, the condition of women has been approaching nearer to equality with men.

"The profoundest knowledge of the laws of the formation of character is indispensable to entitle any one to affirm even that there is any difference, much more what the difference is, between the two sexes, considered as moral and rational beings, and since no one as yet has that knowledge (for there is hardly any subject which, in proportion to its importance, has been so

little studied), no one is thus far entitled to any positive opinion
on the subject.

"The wife is the actual bond-servant of the husband, no
less so, as far as legal obligation goes, than slaves commonly so-
called. She vows a life-long obedience to him at the altar, and
is held to it all through life by law. Casuists may say that the
obligation of obedience stops short of participation in crime, but
it certainly extends to everything else. She can do no act whatever but by his permission—at least, tacit. She can acquire
no property but for him; the instant it becomes hers, even if by
inheritance, it becomes *ipso facto* his. In this respect the
wife's position, under the Common Law of England, is worse
than that of slaves in many countries. By the Roman law, for
example, a slave might have his *peculium*, which, to a certain
extent, the law guaranteed to him for his exclusive use. The
higher classes in this country have given an analogous advantage
to their women through special contracts setting aside the law,
by conditions of pin-money, etc., since, parental feelings being
stronger than the class feelings of their own sex, a father generally prefers his own daughter to a son-in-law, who is a stranger
to him. By means of settlements, the rich usually contrive to
withdraw the whole or part of the inherited property of the
wife from the absolute control of the husband, but they do not
succeed in keeping it under her own control; the utmost they
can do only prevents the husband from squandering it, at the
same time debarring the rightful owner from its use. The
property is out of the reach of both, and as to the income derived from it, the form of settlement most favorable to the wife
(that called 'to her separate use') only precludes the husband
from receiving it instead of her. It must pass through her
hands; but if he takes it from her by personal violence as soon

as she receives it, he can neither be punished nor compelled to restitution."

Enlightened England! Such .is the law. Is it, on the whole, a whit better among our enlightened selves?

In the *Westminster Review* occurs the following mortifying acknowledgment of injustice towards women. It is true enough to make the ears of legislators tingle :—

" This is the wife's *status* with respect to her individual interest, and her *status* in regard to her children is of a piece with it. They are called in law the husband's children, and he alone has legal right over them. The wife can do nothing in relation to them, except by delegation from him, and, even after his death, she does not become their guardian unless she has been appointed so by him.

" The natural sequence and corollary from the state of things here described would be, that since a woman's whole comfort and happiness in life ' depend on her finding a good master, she should be allowed to change, again and again, until she finds one.' "

Here is the opinion of another English thinker, who fully comprehends a problem that political demagogues neither wish to study or understand :—

Mr. Herbert Spencer, speaking of the rights of women, says:—" Three positions only are open to us. It may be said that women have no rights at all; that their rights are not so great as those of men ; or that they are equal to those of men.

" Whoever maintains the first of those dogmas, that women have no rights at all, must show that the Creator intended women to be wholly at the mercy of men—their happiness, their their liberties, their lives, at men's disposal; or, in other words, that they were meant to be treated as creatures of an inferior order. Few will have the hardihood to assert this.

"From the second proposition, that the rights of women are not so great as those of men, there immediately arise such queries as: If they are not so great, by how much are they less? What is the exact ratio between the legitimate claims of the two sexes? How shall we tell which rights are common to both, and where those of the male exceed those of the female? Who can show us a scale that will serve for the apportionment? Or, putting the question practically, it is required to determine, by some logical method, whether the Turk is justified in plunging an offending Circassian into the Bosphorus? Whether the rights of women were violated by the Athenian law, which allowed a citizen, under certain circumstances, to sell his daughter or sister? Whether our own statute, which permits a man to beat his wife in moderation, and to imprison her in any room in his house, is morally defensible? Whether it is equitable that a married woman should be incapable of holding property? Whether a husband may justly take possession of his wife's earnings against her will, as our law allows him to do? —and so forth. These, and a multitude of similar problems, present themselves for solution.

"In this connection it is also curious to contemplate that the only things which women are ordinarily excluded from doing, are just those things which they have proved themselves best able to do. There is no law or custom in force to prevent a woman from writing plays like Shakespeare, or operas like Mozart, but there are laws and customs to prevent them from embracing a military or political career, and Joan of Arc and Queen Elizabeth are historical characters."

Profoundly impressed with the importance of manfully aiding and assisting in the great revolution that is to be ultimately achieved, we offer no apology for strengthening our position from any available source. Another transatlantic view of the women-question, as it is called, here introduced, is too sound and logical not to gain the approval of reasonable men :

" Whoso urges the mental inferiority of women, in bar to their claim to equal rights with men, may be met in various ways. In the first place, the alleged fact may be disputed. A defender of her sex might name many whose achievements in government, in science, in literature, and in art, have obtained no small share of renown. Powerful and sagacious queens the world has seen in plenty, from Zenobia down to the Empresses Catherine and Maria Theresa. In the exact sciences, Mrs. Somerville, Miss Herschel, and Miss Zornlin have gained applause ; in political economy, Miss Martineau ; in general philosophy, Madame de Stael; in politics, Madame Roland. Poetry has its Tighes, its Hemanses, its Landons, its Brownings; the drama, its Joanna Baillies ; and fiction, its Austens, Bremers, Gores, Dudevants, etc., without end. In sculpture, fame has been acquired by a princess ; a picture like ' The Momentous Question ' is tolerable proof of female capacity for painting ; and, on the stage, it is certain that women are on a level with men, if they do not even bear away the palm. Joining to such facts the important consideration, that women have always been, and are still, placed at a disadvantage in every department of learning, thought, or skill—seeing that they are not admissible to the academies and universities in which men get their training ; that the kind of life they have to look forward to does not present so great a range of ambitions; that they are rarely exposed to that most powerful of all stimulants—necessity; that the education custom dictates for them is one that leaves uncultivated many of the higher faculties ; and that the prejudice against blue-stockings, hitherto so prevalent amongst men, has greatly tended to deter women from the pursuit of literary honors : adding these considerations to the above facts, we shall see good reason for thinking that the alleged inferiority of the femine mind is by no means self-evident.

"If we contrast the literary and artistic works of women with those of men in modern days, we shall find that their inferiority resolves itself into one, but still a most material defect, namely, 'a deficiency of originality.' They do not, indeed, exhibit a total want of it, for no production of mind of substantive value can do so; but they have not up to the present been marked 'by any of those great and luminous new ideas which form an era in thought, nor those fundamentally new conceptions in art, which open a vista of possible effects not before thought of, and found a new school.' Their compositions are mostly based on the existing fund of thought, and their creations do not deviate widely from existing types; but in point of execution, in the treatment of details, and in perfection of style, their works are quite on a par with those of their male rivals.

"They are deprived of all the advantages, and most of the motives, which men possess for acquiring even a decent amount of systematic education; and if we turn from philosophy and science to literature, in the narrow sense of the term, there are other obvious reasons why women's productions are, in general conception and in their leading features, more or less imitations of those of men."

Finally, the signs of the times plainly indicate the success of importunate petitioners and aspirants for equal rights. Pioneers and public agitators in the cause of woman's emancipation are indomitable and irrepressible. Concessions are slowly made of unimportant places to their management, which have been singularly well sustained, to the mortification of those who are fighting windmills. More they will have.

What! if a few women should be sent to the legislature, or

to Congress, they would have too much self-respect to have anything to do with any rings but diamond rings, nor would they disgrace themselves by entering into combinations to defraud the Government, foist their imbecile relatives into office, or vote to raise their own pay at the expense of the people already overburdened by excessive taxation.

When women vote, respectable men will be elected over rascals, swindlers, defaulters, and demoralized politicians who are a curse to the country.

**THE END.**

www.ingramcontent.com/pod-product-compliance
Lightning Source LLC
Chambersburg PA
CBHW031810270326
41932CB00008B/369